# Nanoengineering: Science and Technology

# Nanoengineering: Science and Technology

Edited by
**Josiah Jackson**

🖻 Larsen & Keller
www.larsen-keller.com

Nanoengineering: Science and Technology
Edited by Josiah Jackson
ISBN: 978-1-63549-193-7 (Hardback)

## ⊟ Larsen & Keller

Published by Larsen and Keller Education,
5 Penn Plaza,
19th Floor,
New York, NY 10001, USA

### Cataloging-in-Publication Data

Nanoengineering : science and technology / edited by Josiah Jackson.
   p. cm.
Includes bibliographical references and index.
ISBN 978-1-63549-193-7
1. Nanotechnology. 2. Nanoscience. 3. Engineering. I. Jackson, Josiah.
T174.7 .N36 2017
620.5--dc23

This book contains information obtained from authentic and highly regarded sources. All chapters are published with permission under the Creative Commons Attribution Share Alike License or equivalent. A wide variety of references are listed. Permissions and sources are indicated; for detailed attributions, please refer to the permissions page. Reasonable efforts have been made to publish reliable data and information, but the authors, editors and publisher cannot assume any responsibility for the vailidity of all materials or the consequences of their use.

Trademark Notice: All trademarks used herein are the property of their respective owners. The use of any trademark in this text does not vest in the author or publisher any trademark ownership rights in such trademarks, nor does the use of such trademarks imply any affiliation with or endorsement of this book by such owners.

The publisher's policy is to use permanent paper from mills that operate a sustainable forestry policy. Furthermore, the publisher ensures that the text paper and cover boards used have met acceptable environmental accreditation standards.

Printed and bound in the United States of America.

For more information regarding Larsen and Keller Education and its products, please visit the publisher's website www.larsen-keller.com

# Table of Contents

# Preface

Practicing engineering at a nano scale is referred to as nanoengineering. It is a new and rapidly evolving field and its forum as the engineering aspects of nanomaterials. Nanoengineering is seen to be of great help in the fields of pharmaceuticals, genetic engineering and nanotechnology. This book unfolds the innovative aspects of nanoengineering which will be crucial for the holistic understanding of the subject matter. The topics included in it on this important subject are of utmost significance and bound to provide incredible insights to readers. Different approaches, evaluations and methodologies and advanced studies of the field have been included in it. In this text, constant effort has been made to make the understanding of the difficult concepts of nanoengineering as easy and informative as possible, for the readers. It will serve as a resource guide for students and experts alike and contribute to the growth of the discipline.

A foreword of all chapters of the book is provided below:

**Chapter 1** - Nanoengineering is a branch of engineering that deals with the nanoscopic scale. Some of the techniques involved in nanoengineering are scanning tunneling microscope and molecular self-assembly. The chapter will provide an integrated understanding of nanoengineering.; **Chapter 2** - Nanotechnology is the study of matter; the scale of studying matter is on an atomic, molecular and supramolecular scale. Some of the aspects of nanotechnology that have been discussed in this text are nanomechanics, molecular scale electronics, impact of nanotechnology, regulation of nanotechnology and societal impact of nanotechnology. This section is an overview of the subject matter incorporating all the major aspects of nanotechnology.; **Chapter 3** - The essential elements of nanoengineering are nanofluids, nanomaterials, nanoparticles, nanometers, carbon nanotubes and nanosensors. Nanofluids are fluids that contain nanometer-sized particles whereas nanoparticles are materials of size between 1-100 nanoscale. The topics discussed in the chapter are of great importance to broaden the existing knowledge of nanoengineering.; **Chapter 4** - The techniques used for nanoengineering are scanning tunneling microscope and molecular self-assembly. Scanning tunneling microscope is an instrument that is used to scan surfaces at the atomic level whereas the process by which molecules arrange themselves without any guidance is termed as molecular self-assembly. This text discusses the methods of nanoengineering in a critical manner providing key analysis to the subject matter.; **Chapter 5** - This section discusses the applications and designs of nanoengineering. The medical function of nanotechnology is nanomedicine whereas the study of nanomaterials is nanotoxicology. The other applications and designs that have been discussed are nanoelectronics, green nanotechnology, DNA nanotechnology, nanoremediation and nanofiltration. The topics elaborated in the chapter will help in gaining a better perspective about the applications of nanoengineering.; **Chapter 6** - Nanoengineering is a vast subject that has a number of allied fields that have been thoroughly discussed in this section. Some of the fields explained in the following text are nanophotonics, materials science, molecular engineering, tissue engineering, ceramic engineering and nanometrology. The study of the behavior of light on the scale of nanometer is known as nanohotonics and the discovery and designing of new materials is referred to as materials science and engineering. In order to completely understand nanoengineering, it is necessary to understand the fields allied to it.; **Chapter 7** - Nanoengineering has evolved over a period of years. It traces the growth of the ideas and concepts falling under the broad category of nanoengineering. The content within the chapter helps the reader in developing an in-depth understanding of the history of nanoengineering and nanotechnology.

At the end, I would like to thank all the people associated with this book devoting their precious time and providing their valuable contributions to this book. I would also like to express my gratitude to my fellow colleagues who encouraged me throughout the process.

**Editor**

# Introduction to Nanoengineering

Nanoengineering is a branch of engineering that deals with the nanoscopic scale. Some of the techniques involved in nanoengineering are scanning tunneling microscope and molecular self-assembly. The chapter will provide an integrated understanding of nanoengineering.

Nanoengineering is the practice of engineering on the nanoscale. It derives its name from the nanometre, a unit of measurement equalling one billionth of a meter.

Nanoengineering is largely a synonym for nanotechnology, but emphasizes the engineering rather than the pure science aspects of the field.

## Degree Programs

The first nanoengineering program in the world was started at the University of Toronto within the Engineering Science program as one of the Options of study in the final years. In 2003, the Lund Institute of Technology started a program in Nanoengineering. In 2004, the College of Nanoscale Science and Engineering at SUNY Polytechnic Institute was established on the campus of the University at Albany. In 2005, the University of Waterloo established a unique program which offers a full degree in Nanotechnology Engineering. Louisiana Tech University started the first program in the U.S. in 2005. In 2006 the University of Duisburg-Essen started a Bachelor and a Master program Nano-Engineering. The University of California, San Diego followed shortly thereafter in 2007 with its own department of Nanoengineering. In 2009, the University of Toronto began offering all Options of study in Engineering Science as degrees, bringing the second nanoengineering degree to Canada. DTU Nanotech - the Department of Micro- and Nanotechnology - is a department at the Technical University of Denmark established in 1990.

In 2013, Wayne State University began offering a Nanoengineering Undergraduate Certificate Program, which is funded by a Nanoengineering Undergraduate Education (NUE) grant from the National Science Foundation. The primary goal is to offer specialized undergraduate training in nanotechnology. Other goals are: 1) to teach emerging technologies at the undergraduate level, 2) to train a new adaptive workforce, and 3) to retrain working engineers and professionals.

## Techniques

- Scanning tunneling microscope (STM) - Can be used to both image, and to manipulate structures as small as a single atom.

- Molecular self-assembly - Arbitrary sequences of DNA can now be synthesized cheaply in bulk, and used to create custom proteins or regular patterns of amino acids. Similarly, DNA strands can bind to other DNA strands, allowing simple structures to be created.

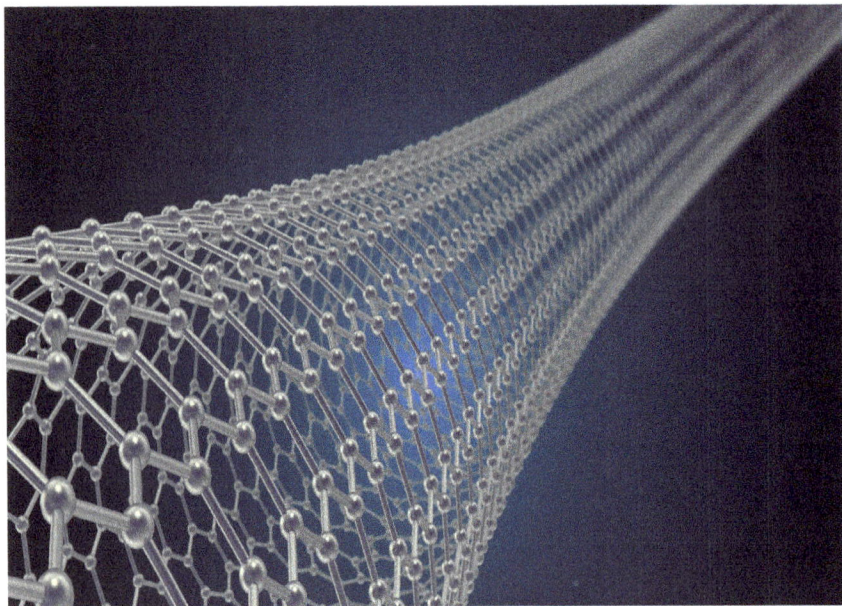

# Nanotechnology: An Overview

Nanotechnology is the study of matter; the scale of studying matter is on an atomic, molecular and supramolecular scale. Some of the aspects of nanotechnology that have been discussed in this text are nanomechanics, molecular scale electronics, impact of nanotechnology, regulation of nanotechnology and societal impact of nanotechnology. This section is an overview of the subject matter incorporating all the major aspects of nanotechnology.

## Nanotechnology

Nanotechnology ("nanotech") is manipulation of matter on an atomic, molecular, and supramolecular scale. The earliest, widespread description of nanotechnology referred to the particular technological goal of precisely manipulating atoms and molecules for fabrication of macroscale products, also now referred to as molecular nanotechnology. A more generalized description of nanotechnology was subsequently established by the National Nanotechnology Initiative, which defines nanotechnology as the manipulation of matter with at least one dimension sized from 1 to 100 nanometers. This definition reflects the fact that quantum mechanical effects are important at this quantum-realm scale, and so the definition shifted from a particular technological goal to a research category inclusive of all types of research and technologies that deal with the special properties of matter which occur below the given size threshold. It is therefore common to see the plural form "nanotechnologies" as well as "nanoscale technologies" to refer to the broad range of research and applications whose common trait is size. Because of the variety of potential applications (including industrial and military), governments have invested billions of dollars in nanotechnology research. Until 2012, through its National Nanotechnology Initiative, the USA has invested 3.7 billion dollars, the European Union has invested 1.2 billion and Japan 750 million dollars.

Nanotechnology as defined by size is naturally very broad, including fields of science as diverse as surface science, organic chemistry, molecular biology, semiconductor physics, microfabrication, molecular engineering, etc. The associated research and applications are equally diverse, ranging from extensions of conventional device physics to completely new approaches based upon molecular self-assembly, from developing new materials with dimensions on the nanoscale to direct control of matter on the atomic scale.

Scientists currently debate the future implications of nanotechnology. Nanotechnology may be able to create many new materials and devices with a vast range of applications, such as in nanomedicine, nanoelectronics, biomaterials energy production, and consumer products. On the other hand, nanotechnology raises many of the same issues as any new technology, including concerns about the toxicity and environmental impact of nanomaterials, and their potential effects on global economics, as well as speculation about various doomsday scenarios. These concerns have led to a debate among advocacy groups and governments on whether special regulation of nanotechnology is warranted.

## Origins

The concepts that seeded nanotechnology were first discussed in 1959 by renowned physicist Richard Feynman in his talk *There's Plenty of Room at the Bottom*, in which he described the possibility of synthesis via direct manipulation of atoms. The term "nano-technology" was first used by Norio Taniguchi in 1974, though it was not widely known.

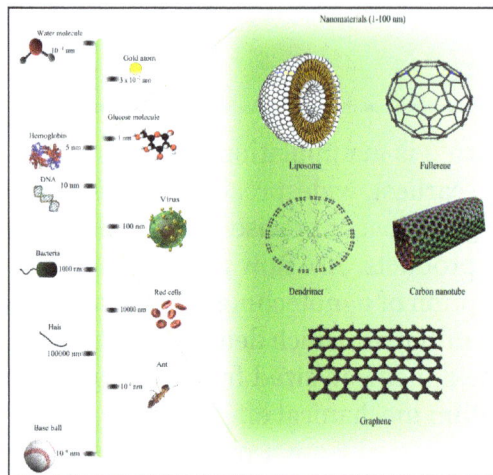

Comparison of Nanomaterials Sizes

Inspired by Feynman's concepts, K. Eric Drexler used the term "nanotechnology" in his 1986 book *Engines of Creation: The Coming Era of Nanotechnology*, which proposed the idea of a nanoscale "assembler" which would be able to build a copy of itself and of other items of arbitrary complexity with atomic control. Also in 1986, Drexler co-founded The Foresight Institute (with which he is no longer affiliated) to help increase public awareness and understanding of nanotechnology concepts and implications.

Thus, emergence of nanotechnology as a field in the 1980s occurred through convergence of Drexler's theoretical and public work, which developed and popularized a conceptual framework for nanotechnology, and high-visibility experimental advances that drew additional wide-scale attention to the prospects of atomic control of matter. In the 1980s, two major breakthroughs sparked the growth of nanotechnology in modern era.

First, the invention of the scanning tunneling microscope in 1981 which provided unprecedented visualization of individual atoms and bonds, and was successfully used to manipulate individual atoms in 1989. The microscope's developers Gerd Binnig and

Heinrich Rohrer at IBM Zurich Research Laboratory received a Nobel Prize in Physics in 1986. Binnig, Quate and Gerber also invented the analogous atomic force microscope that year.

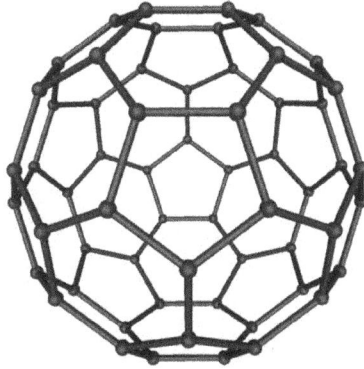

Buckminsterfullerene $C_{60}$, also known as the buckyball, is a representative member of the carbon structures known as fullerenes. Members of the fullerene family are a major subject of research falling under the nanotechnology umbrella.

Second, Fullerenes were discovered in 1985 by Harry Kroto, Richard Smalley, and Robert Curl, who together won the 1996 Nobel Prize in Chemistry. $C_{60}$ was not initially described as nanotechnology; the term was used regarding subsequent work with related graphene tubes (called carbon nanotubes and sometimes called Bucky tubes) which suggested potential applications for nanoscale electronics and devices.

In the early 2000s, the field garnered increased scientific, political, and commercial attention that led to both controversy and progress. Controversies emerged regarding the definitions and potential implications of nanotechnologies, exemplified by the Royal Society's report on nanotechnology. Challenges were raised regarding the feasibility of applications envisioned by advocates of molecular nanotechnology, which culminated in a public debate between Drexler and Smalley in 2001 and 2003.

Meanwhile, commercialization of products based on advancements in nanoscale technologies began emerging. These products are limited to bulk applications of nanomaterials and do not involve atomic control of matter. Some examples include the Silver Nano platform for using silver nanoparticles as an antibacterial agent, nanoparticle-based transparent sunscreens, carbon fiber strengthening using silica nanoparticles, and carbon nanotubes for stain-resistant textiles.

Governments moved to promote and fund research into nanotechnology, such as in the U.S. with the National Nanotechnology Initiative, which formalized a size-based definition of nanotechnology and established funding for research on the nanoscale, and in Europe via the European Framework Programmes for Research and Technological Development.

By the mid-2000s new and serious scientific attention began to flourish. Projects emerged to produce nanotechnology roadmaps which center on atomically precise manipulation of matter and discuss existing and projected capabilities, goals, and applications.

## Fundamental Concepts

Nanotechnology is the engineering of functional systems at the molecular scale. This covers both current work and concepts that are more advanced. In its original sense, nanotechnology refers to the projected ability to construct items from the bottom up, using techniques and tools being developed today to make complete, high performance products.

One nanometer (nm) is one billionth, or $10^{-9}$, of a meter. By comparison, typical carbon-carbon bond lengths, or the spacing between these atoms in a molecule, are in the range 0.12–0.15 nm, and a DNA double-helix has a diameter around 2 nm. On the other hand, the smallest cellular life-forms, the bacteria of the genus Mycoplasma, are around 200 nm in length. By convention, nanotechnology is taken as the scale range 1 to 100 nm following the definition used by the National Nanotechnology Initiative in the US. The lower limit is set by the size of atoms (hydrogen has the smallest atoms, which are approximately a quarter of a nm diameter) since nanotechnology must build its devices from atoms and molecules. The upper limit is more or less arbitrary but is around the size below which phenomena not observed in larger structures start to become apparent and can be made use of in the nano device. These new phenomena make nanotechnology distinct from devices which are merely miniaturised versions of an equivalent macroscopic device; such devices are on a larger scale and come under the description of microtechnology.

To put that scale in another context, the comparative size of a nanometer to a meter is the same as that of a marble to the size of the earth. Or another way of putting it: a nanometer is the amount an average man's beard grows in the time it takes him to raise the razor to his face.

Two main approaches are used in nanotechnology. In the "bottom-up" approach, materials and devices are built from molecular components which assemble themselves chemically by principles of molecular recognition. In the "top-down" approach, nano-objects are constructed from larger entities without atomic-level control.

Areas of physics such as nanoelectronics, nanomechanics, nanophotonics and nano-ionics have evolved during the last few decades to provide a basic scientific foundation of nanotechnology.

## Larger to Smaller: A Materials Perspective

Several phenomena become pronounced as the size of the system decreases. These include statistical mechanical effects, as well as quantum mechanical effects, for example the "quantum size effect" where the electronic properties of solids are altered with great reductions in particle size. This effect does not come into play by going from macro to micro dimensions. However, quantum effects can become significant when the nanometer size range is reached, typically at distances of 100 nanometers or less, the so-called quantum realm. Additionally, a number of physical (mechanical, electrical, optical, etc.) properties change when compared to macroscopic systems. One example is the increase in surface area to volume ratio altering mechanical, thermal and catalytic properties of materials. Diffusion and reactions at nanoscale, nanostructures materials and nanodevices with fast ion transport are generally referred to nanoionics. *Mechanical* properties of nanosystems are of interest in the nanomechanics research. The catalytic activity of nanomaterials also opens potential risks in their interaction with biomaterials.

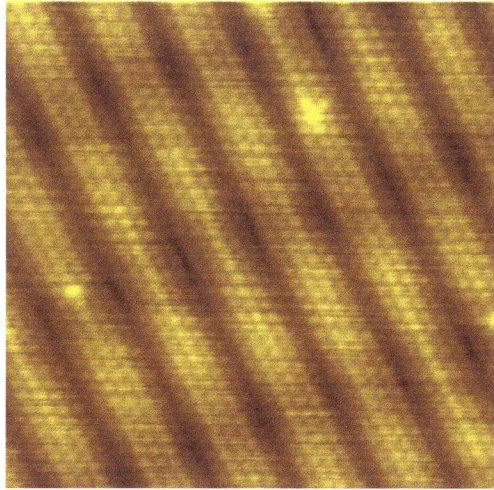

Image of reconstruction on a clean Gold(100) surface, as visualized using scanning tunneling microscopy. The positions of the individual atoms composing the surface are visible.

Materials reduced to the nanoscale can show different properties compared to what they exhibit on a macroscale, enabling unique applications. For instance, opaque substances can become transparent (copper); stable materials can turn combustible (aluminium); insoluble materials may become soluble (gold). A material such as gold, which is chemically inert at normal scales, can serve as a potent chemical catalyst at nanoscales. Much of the fascination with nanotechnology stems from these quantum and surface phenomena that matter exhibits at the nanoscale.

## Simple to Complex: A Molecular Perspective

Modern synthetic chemistry has reached the point where it is possible to prepare small molecules to almost any structure. These methods are used today to manufacture a wide variety of useful chemicals such as pharmaceuticals or commercial polymers. This ability raises the question of extending this kind of control to the next-larger level, seeking methods to assemble these single molecules into supramolecular assemblies consisting of many molecules arranged in a well defined manner.

These approaches utilize the concepts of molecular self-assembly and/or supramolecular chemistry to automatically arrange themselves into some useful conformation through a bottom-up approach. The concept of molecular recognition is especially important: molecules can be designed so that a specific configuration or arrangement is favored due to non-covalent intermolecular forces. The Watson–Crick basepairing rules are a direct result of this, as is the specificity of an enzyme being targeted to a single substrate, or the specific folding of the protein itself. Thus, two or more components can be designed to be complementary and mutually attractive so that they make a more complex and useful whole.

Such bottom-up approaches should be capable of producing devices in parallel and be much cheaper than top-down methods, but could potentially be overwhelmed as the size and complexity of the desired assembly increases. Most useful structures require complex and thermodynamically unlikely arrangements of atoms. Nevertheless, there are many examples of self-assembly based on molecular recognition in biology, most notably Watson–Crick basepairing and enzyme-substrate interactions. The challenge for nanotechnology is whether these principles can be used to engineer new constructs in addition to natural ones.

## Molecular Nanotechnology: A Long-term View

Molecular nanotechnology, sometimes called molecular manufacturing, describes engineered nanosystems (nanoscale machines) operating on the molecular scale. Molecular nanotechnology is especially associated with the molecular assembler, a machine that can produce a desired structure or device atom-by-atom using the principles of mechanosynthesis. Manufacturing in the context of productive nanosystems is not related to, and should be clearly distinguished from, the conventional technologies used to manufacture nanomaterials such as carbon nanotubes and nanoparticles.

When the term "nanotechnology" was independently coined and popularized by Eric Drexler (who at the time was unaware of an earlier usage by Norio Taniguchi) it referred to a future manufacturing technology based on molecular machine systems. The premise was that molecular scale biological analogies of traditional machine components demonstrated molecular machines were possible: by the countless examples found in biology, it is known that sophisticated, stochastically optimised biological machines can be produced.

It is hoped that developments in nanotechnology will make possible their construction by some other means, perhaps using biomimetic principles. However, Drexler and other researchers have proposed that advanced nanotechnology, although perhaps initially implemented by biomimetic means, ultimately could be based on mechanical engineering principles, namely, a manufacturing technology based on the mechanical functionality of these components (such as gears, bearings, motors, and structural members) that would enable programmable, positional assembly to atomic specification. The physics and engineering performance of exemplar designs were analyzed in Drexler's book *Nanosystems*.

In general it is very difficult to assemble devices on the atomic scale, as one has to position atoms on other atoms of comparable size and stickiness. Another view, put forth by Carlo Montemagno, is that future nanosystems will be hybrids of silicon technology and biological molecular machines. Richard Smalley argued that mechanosynthesis are impossible due to the difficulties in mechanically manipulating individual molecules.

This led to an exchange of letters in the ACS publication Chemical & Engineering News in 2003. Though biology clearly demonstrates that molecular machine systems are possible, non-biological molecular machines are today only in their infancy. Leaders in research on non-biological molecular machines are Dr. Alex Zettl and his colleagues at Lawrence Berkeley Laboratories and UC Berkeley. They have constructed at least three distinct molecular devices whose motion is controlled from the desktop with changing voltage: a nanotube nanomotor, a molecular actuator, and a nanoelectromechanical relaxation oscillator.

An experiment indicating that positional molecular assembly is possible was performed by Ho and Lee at Cornell University in 1999. They used a scanning tunneling microscope to move an individual carbon monoxide molecule (CO) to an individual iron atom (Fe) sitting on a flat silver crystal, and chemically bound the CO to the Fe by applying a voltage.

## Current Research

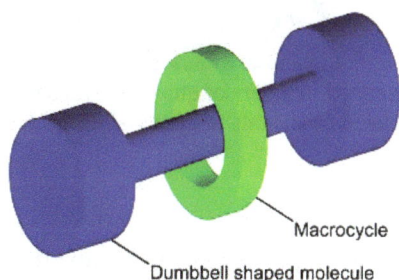

Graphical representation of a rotaxane, useful as a molecular switch.

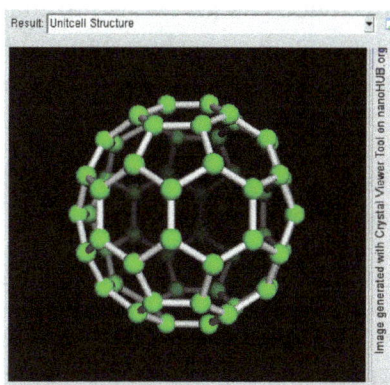

Rotating view of $C_{60}$, one kind of fullerene.

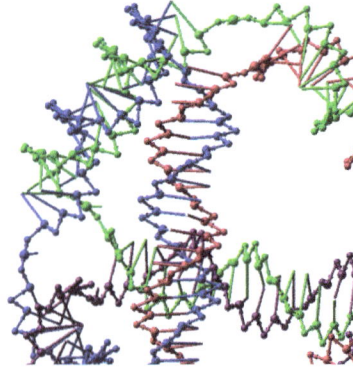

This DNA tetrahedron is an artificially designed nanostructure of the type made in the field of DNA nanotechnology. Each edge of the tetrahedron is a 20 base pair DNA double helix, and each vertex is a three-arm junction.

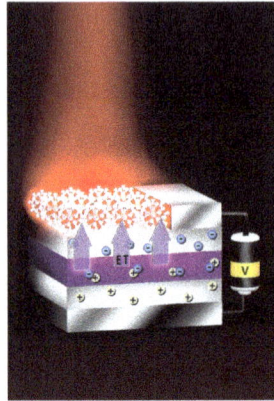

This device transfers energy from nano-thin layers of quantum wells to nanocrystals above them, causing the nanocrystals to emit visible light.

## Nanomaterials

The nanomaterials field includes subfields which develop or study materials having unique properties arising from their nanoscale dimensions.

- Interface and colloid science has given rise to many materials which may be useful in nanotechnology, such as carbon nanotubes and other fullerenes, and various nanoparticles and nanorods. Nanomaterials with fast ion transport are related also to nanoionics and nanoelectronics.

- Nanoscale materials can also be used for bulk applications; most present commercial applications of nanotechnology are of this flavor.

- Progress has been made in using these materials for medical applications.

- Nanoscale materials such as nanopillars are sometimes used in solar cells which combats the cost of traditional Silicon solar cells.

- Development of applications incorporating semiconductor nanoparticles to be used in the next generation of products, such as display technology, lighting, solar cells and biological imaging.

- Recent application of nanomaterials include a range of biomedical applications, such as tissue engineering, drug delivery, and biosensors.

## Bottom-up Approaches

These seek to arrange smaller components into more complex assemblies.

- DNA nanotechnology utilizes the specificity of Watson–Crick basepairing to construct well-defined structures out of DNA and other nucleic acids.

- Approaches from the field of "classical" chemical synthesis (Inorganic and organic synthesis) also aim at designing molecules with well-defined shape (e.g. bis-peptides).

- More generally, molecular self-assembly seeks to use concepts of supramolecular chemistry, and molecular recognition in particular, to cause single-molecule components to automatically arrange themselves into some useful conformation.

- Atomic force microscope tips can be used as a nanoscale "write head" to deposit a chemical upon a surface in a desired pattern in a process called dip pen nanolithography. This technique fits into the larger subfield of nanolithography.

## Top-down Approaches

These seek to create smaller devices by using larger ones to direct their assembly.

- Many technologies that descended from conventional solid-state silicon methods for fabricating microprocessors are now capable of creating features smaller than 100 nm, falling under the definition of nanotechnology. Giant magnetoresistance-based hard drives already on the market fit this description, as do atomic layer deposition (ALD) techniques. Peter Grünberg and Albert Fert received the Nobel Prize in Physics in 2007 for their discovery of Giant magnetoresistance and contributions to the field of spintronics.

- Solid-state techniques can also be used to create devices known as nanoelectromechanical systems or NEMS, which are related to microelectromechanical systems or MEMS.

- Focused ion beams can directly remove material, or even deposit material when suitable precursor gasses are applied at the same time. For example, this technique is used routinely to create sub-100 nm sections of material for analysis in Transmission electron microscopy.

- Atomic force microscope tips can be used as a nanoscale "write head" to deposit a resist, which is then followed by an etching process to remove material in a top-down method.

## Functional Approaches

These seek to develop components of a desired functionality without regard to how they might be assembled.

- Magnetic assembly for the synthesis of anisotropic superparamagnetic materials such as recently presented magnetic nanochains.

- Molecular scale electronics seeks to develop molecules with useful electronic properties. These could then be used as single-molecule components in a nano-electronic device. For an example rotaxane.

- Synthetic chemical methods can also be used to create synthetic molecular motors, such as in a so-called nanocar.

## Biomimetic Approaches

- Bionics or biomimicry seeks to apply biological methods and systems found in nature, to the study and design of engineering systems and modern technology. Biomineralization is one example of the systems studied.

- Bionanotechnology is the use of biomolecules for applications in nanotechnology, including use of viruses and lipid assemblies. Nanocellulose is a potential bulk-scale application.

## Speculative

These subfields seek to anticipate what inventions nanotechnology might yield, or attempt to propose an agenda along which inquiry might progress. These often take a big-picture view of nanotechnology, with more emphasis on its societal implications than the details of how such inventions could actually be created.

- Molecular nanotechnology is a proposed approach which involves manipulating single molecules in finely controlled, deterministic ways. This is more theoretical than the other subfields, and many of its proposed techniques are beyond current capabilities.

- Nanorobotics centers on self-sufficient machines of some functionality operating at the nanoscale. There are hopes for applying nanorobots in medicine, but it may not be easy to do such a thing because of several drawbacks of such devices. Nevertheless, progress on innovative materials and methodologies has been demonstrated with some patents granted about new nanomanufacturing devices for future commercial applications, which also progressively helps in the development towards nanorobots with the use of embedded nanobioelectronics concepts.

- Productive nanosystems are "systems of nanosystems" which will be complex nanosystems that produce atomically precise parts for other nanosystems, not necessarily using novel nanoscale-emergent properties, but well-understood fundamentals of manufacturing. Because of the discrete (i.e. atomic) nature of matter and the possibility of exponential growth, this stage is seen as the basis of another industrial revolution. Mihail Roco, one of the architects of the USA's National Nanotechnology Initiative, has proposed four states of nanotechnology that seem to parallel the technical progress of the Industrial Revolution, progressing from passive nanostructures to active nanodevices to complex nanomachines and ultimately to productive nanosystems.

- Programmable matter seeks to design materials whose properties can be easily, reversibly and externally controlled though a fusion of information science and materials science.

- Due to the popularity and media exposure of the term nanotechnology, the words picotechnology and femtotechnology have been coined in analogy to it, although these are only used rarely and informally.

## Dimensionality in Nanomaterials

Nanomaterials can be classified in 0D, 1D, 2D and 3D nanomaterials. The dimensionality play a major role in determining the characteristic of nanomaterials including physical, chemical and biological characteristics. With the decrease in dimensionality, an increase in surface-to-volume ratio is observed. This indicate that smaller dimensional nanomaterials have higher surface area compared to 3D nanomaterials. Recently, two dimensional (2D) nanomaterials are extensively investigated for electronic, biomedical, drug delivery and biosensor applications.

## Tools and Techniques

There are several important modern developments. The atomic force microscope (AFM) and the Scanning Tunneling Microscope (STM) are two early versions of scanning probes that launched nanotechnology. There are other types of scanning probe microscopy. Although conceptually similar to the scanning confocal microscope developed by Marvin Minsky in 1961 and the scanning acoustic microscope (SAM) developed by Calvin Quate and coworkers in the 1970s, newer scanning probe microscopes have much higher resolution, since they are not limited by the wavelength of sound or light.

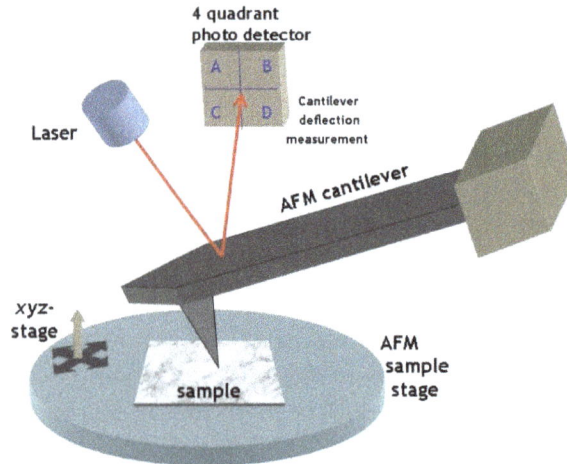

Typical AFM setup. A microfabricated cantilever with a sharp tip is deflected by features on a sample surface, much like in a phonograph but on a much smaller scale. A laser beam reflects off the backside of the cantilever into a set of photodetectors, allowing the deflection to be measured and assembled into an image of the surface.

The tip of a scanning probe can also be used to manipulate nanostructures (a process called positional assembly). Feature-oriented scanning methodology may be a promising way to implement these nanomanipulations in automatic mode. However, this is still a slow process because of low scanning velocity of the microscope.

Various techniques of nanolithography such as optical lithography, X-ray lithography dip pen nanolithography, electron beam lithography or nanoimprint lithography were also developed. Lithography is a top-down fabrication technique where a bulk material is reduced in size to nanoscale pattern.

Another group of nanotechnological techniques include those used for fabrication of nanotubes and nanowires, those used in semiconductor fabrication such as deep ultraviolet lithography, electron beam lithography, focused ion beam machining, nanoimprint lithography, atomic layer deposition, and molecular vapor deposition, and further including molecular self-assembly techniques such as those employing di-block copolymers. The precursors of these techniques preceded the nanotech era, and are extensions in the development of scientific advancements rather than techniques which were devised with the sole purpose of creating nanotechnology and which were results of nanotechnology research.

The top-down approach anticipates nanodevices that must be built piece by piece in stages, much as manufactured items are made. Scanning probe microscopy is an important technique both for characterization and synthesis of nanomaterials. Atomic force microscopes and scanning tunneling microscopes can be used to look at surfaces and to move atoms around. By designing different tips for these microscopes, they can be used for carving out structures on surfaces and to help guide self-assembling structures. By using, for example, feature-oriented scanning approach, atoms or molecules can be moved around on a surface with scanning probe microscopy techniques. At present, it is expensive and time-consuming for mass production but very suitable for laboratory experimentation.

In contrast, bottom-up techniques build or grow larger structures atom by atom or molecule by molecule. These techniques include chemical synthesis, self-assembly and positional assembly. Dual polarisation interferometry is one tool suitable for characterisation of self assembled thin films. Another variation of the bottom-up approach is molecular beam epitaxy or MBE. Researchers at Bell Telephone Laboratories like John R. Arthur. Alfred Y. Cho, and Art C. Gossard developed and implemented MBE as a research tool in the late 1960s and 1970s. Samples made by MBE were key to the discovery of the fractional quantum Hall effect for which the 1998 Nobel Prize in Physics was awarded. MBE allows scientists to lay down atomically precise layers of atoms and, in the process, build up complex structures. Important for research on semiconductors, MBE is also widely used to make samples and devices for the newly emerging field of spintronics.

However, new therapeutic products, based on responsive nanomaterials, such as the ultradeformable, stress-sensitive Transfersome vesicles, are under development and already approved for human use in some countries.

## Applications

Nanostructures provide this surface with superhydrophobicity, which lets water droplets roll down the inclined plane.

As of August 21, 2008, the Project on Emerging Nanotechnologies estimates that over 800 manufacturer-identified nanotech products are publicly available, with new ones hitting the market at a pace of 3–4 per week. The project lists all of the products in a publicly accessible online database. Most applications are limited to the use of "first generation" passive nanomaterials which includes titanium dioxide in sunscreen, cosmetics, surface coatings, and some food products; Carbon allotropes used to produce gecko tape; silver in food packaging, clothing, disinfectants and household appliances; zinc oxide in sunscreens and cosmetics, surface coatings, paints and outdoor furniture varnishes; and cerium oxide as a fuel catalyst.

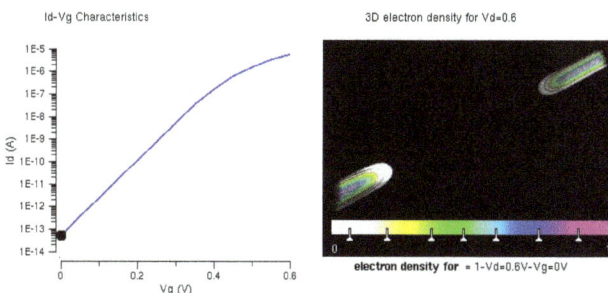

One of the major applications of nanotechnology is in the area of nanoelectronics with MOSFET's being made of small nanowires ~10 nm in length. Here is a simulation of such a nanowire.

Further applications allow tennis balls to last longer, golf balls to fly straighter, and even bowling balls to become more durable and have a harder surface. Trousers and socks have been infused with nanotechnology so that they will last longer and keep people cool in the summer. Bandages are being infused with silver nanoparticles to heal cuts faster. Video game consoles and personal computers may become cheaper, faster, and contain more memory thanks to nanotechnology. Nanotechnology may have the ability to make existing medical applications cheaper and easier to use in places like the general practitioner's office and at home. Cars are being manufactured with nanomaterials so they may need fewer metals and less fuel to operate in the future.

Scientists are now turning to nanotechnology in an attempt to develop diesel engines with cleaner exhaust fumes. Platinum is currently used as the diesel engine catalyst in these engines. The catalyst is what cleans the exhaust fume particles. First a reduction catalyst is employed to take nitrogen atoms from NOx molecules in order to free oxygen. Next the oxidation catalyst oxidizes the hydrocarbons and carbon monoxide to form carbon dioxide and water. Platinum is used in both the reduction and the oxidation catalysts. Using platinum though, is inefficient in that it is expensive and unsustainable. Danish company InnovationsFonden invested DKK 15 million in a search for new catalyst substitutes using nanotechnology. The goal of the project, launched in the autumn of 2014, is to maximize surface area and minimize the amount of material required. Objects tend to minimize their surface energy; two drops of water, for example, will join to form one drop and decrease surface area. If the catalyst's surface area that is exposed to the exhaust fumes is maximized, efficiency of the catalyst is maximized. The team working on this project aims to create nanoparticles that will not merge. Every time the surface is optimized, material is saved. Thus, creating these nanoparticles will increase the effectiveness of the resulting diesel engine catalyst—in turn leading to cleaner exhaust fumes—and will decrease cost. If successful, the team hopes to reduce platinum use by 25%.

Nanotechnology also has a prominent role in the fast developing field of Tissue Engineering. When designing scaffolds, researchers attempt to the mimic the nanoscale features of a Cell's microenvironment to direct its differentiation down a suitable lineage. For example, when creating scaffolds to support the growth of bone, researchers may mimic osteoclast resorption pits.

Researchers have successfully used DNA origami-based nanobots capable of carrying out logic functions to achieve targeted drug delivery in cockroaches. It is said that the computational power of these nanobots can be scaled up to that of a Commodore 64.

## Implications

An area of concern is the effect that industrial-scale manufacturing and use of nanomaterials would have on human health and the environment, as suggested by nanotoxicology research. For these reasons, some groups advocate that nanotechnology be regulated by governments. Others counter that overregulation would stifle scientific research and the development of beneficial innovations. Public health research agencies, such as the National Institute for Occupational Safety and Health are actively conducting research on potential health effects stemming from exposures to nanoparticles.

Some nanoparticle products may have unintended consequences. Researchers have discovered that bacteriostatic silver nanoparticles used in socks to reduce foot odor are being released in the wash. These particles are then flushed into the waste water stream and may destroy bacteria which are critical components of natural ecosystems, farms, and waste treatment processes.

Public deliberations on risk perception in the US and UK carried out by the Center for Nanotechnology in Society found that participants were more positive about nanotechnologies for energy applications than for health applications, with health applications raising moral and ethical dilemmas such as cost and availability.

Experts, including director of the Woodrow Wilson Center's Project on Emerging Nanotechnologies David Rejeski, have testified that successful commercialization depends on adequate oversight, risk research strategy, and public engagement. Berkeley, California is currently the only city in the United States to regulate nanotechnology; Cambridge, Massachusetts in 2008 considered enacting a similar law, but ultimately rejected it. Relevant for both research on and application of nanotechnologies, the insurability of nanotechnology is contested. Without state regulation of nanotechnology, the availability of private insurance for potential damages is seen as necessary to ensure that burdens are not socialised implicitly.

## Health and Environmental Concerns

Nanofibers are used in several areas and in different products, in everything from aircraft wings to tennis rackets. Inhaling airborne nanoparticles and nanofibers may lead to a number of pulmonary diseases, e.g. fibrosis. Researchers have found that when rats breathed in nanoparticles, the particles settled in the brain and lungs, which led to significant increases in biomarkers for inflammation and stress response and that nanoparticles induce skin aging through oxidative stress in hairless mice.

A two-year study at UCLA's School of Public Health found lab mice consuming nano-titanium dioxide showed DNA and chromosome damage to a degree "linked to all the big killers of man, namely cancer, heart disease, neurological disease and aging".

A major study published more recently in Nature Nanotechnology suggests some forms of carbon nanotubes – a poster child for the "nanotechnology revolution" – could be as harmful as asbestos if inhaled in sufficient quantities. Anthony Seaton of the Institute of Occupational Medicine in Edinburgh, Scotland, who contributed to the article on carbon nanotubes said "We know that some of them probably have the potential to cause mesothelioma. So those sorts of materials need to be handled very carefully."

In the absence of specific regulation forthcoming from governments, Paull and Lyons (2008) have called for an exclusion of engineered nanoparticles in food. A newspaper article reports that workers in a paint factory developed serious lung disease and nanoparticles were found in their lungs.

## Regulation

Calls for tighter regulation of nanotechnology have occurred alongside a growing debate related to the human health and safety risks of nanotechnology. There is significant debate about who is responsible for the regulation of nanotechnology. Some regulatory agencies currently cover some nanotechnology products and processes (to varying degrees) – by "bolting on" nanotechnology to existing regulations – there are clear gaps in these regimes. Davies (2008) has proposed a regulatory road map describing steps to deal with these shortcomings.

Stakeholders concerned by the lack of a regulatory framework to assess and control risks associated with the release of nanoparticles and nanotubes have drawn parallels with bovine spongiform encephalopathy ("mad cow" disease), thalidomide, genetically modified food, nuclear energy, reproductive technologies, biotechnology, and asbestosis. Dr. Andrew Maynard, chief science advisor to the Woodrow Wilson Center's Project on Emerging Nanotechnologies, concludes that there is insufficient funding for human health and safety research, and as a result there is currently limited understanding of the human health and safety risks associated with nanotechnology. As a result, some academics have called for stricter application of the precautionary principle, with delayed marketing approval, enhanced labelling and additional safety data development requirements in relation to certain forms of nanotechnology.

The Royal Society report identified a risk of nanoparticles or nanotubes being released during disposal, destruction and recycling, and recommended that "manufacturers of products that fall under extended producer responsibility regimes such as end-of-life regulations publish procedures outlining how these materials will be managed to minimize possible human and environmental exposure" (p. xiii).

The Center for Nanotechnology in Society has found that people respond to nanotechnologies differently, depending on application – with participants in public deliberations more positive about nanotechnologies for energy than health applications – suggesting that any public calls for nano regulations may differ by technology sector.

# Nanomechanics

Nanomechanics is a branch of *nanoscience* studying fundamental *mechanical* (elastic, thermal and kinetic) properties of physical systems at the nanometer scale. Nanomechanics has emerged on the crossroads of classical mechanics, solid-state physics, statistical mechanics, materials science, and quantum chemistry. As an area of nanoscience, nanomechanics provides a scientific foundation of nanotechnology.

Nanomechanics is that branch of nanoscience which deals with the study and application of fundamental mechanical properties of physical systems at the nanoscale, such as elastic, thermal and kinetic material properties.

Often, nanomechanics is viewed as a *branch* of nanotechnology, i.e., an applied area with a focus on the mechanical properties of *engineered* nanostructures and nanosystems (systems with nanoscale components of importance). Examples of the latter include nanoparticles, nanopowders, nanowires, nanorods, nanoribbons, nanotubes, including carbon nanotubes (CNT) and boron nitride nanotubes (BNNTs); nanoshells, nanomebranes, nanocoatings, nanocomposite/nanostructured materials, (fluids with dispersed nanoparticles); nanomotors, etc.

Some of the well-established *fields of nanomechanics* are: nanomaterials, nanotribology (friction, wear and contact mechanics at the nanoscale), nanoelectromechanical systems (NEMS), and nanofluidics.

As a fundamental science, nanomechanics is based on some empirical principles (basic observations), namely general mechanics principles and specific principles arising from the smallness of physical sizes of the object of study.

General mechanics principles include:

- Energy and momentum conservation principles
- Variational Hamilton's principle
- Symmetry principles

Due to smallness of the studied object, nanomechanics also accounts for:

- *Discreteness* of the object, whose size is comparable with the interatomic distances
- Plurality, but *finiteness*, of degrees of freedom in the object
- Importance of *thermal fluctuations*
- Importance of *entropic effects*
- Importance of *quantum effects*

These principles serve to provide a basic insight into novel mechanical properties of nanometer objects. Novelty is understood in the sense that these properties are not present in similar macroscale objects or much different from the properties of those (e.g., nanorods vs. usual macroscopic beam structures). In particular, smallness of the subject itself gives rise to various surface effects determined by higher surface-to-volume ratio of nanostructures, and thus affects mechanoenergetic and thermal properties (melting point, heat capacitance, etc.) of nanostructures. Discreteness serves a fundamental reason, for instance, for the dispersion of mechanical waves in solids, and some special behavior of basic elastomechanics solutions at small scales. Plurality of degrees of freedom and the rise of thermal fluctuations are the reasons for thermal tunneling of nanoparticles through potential barriers, as well as for the cross-diffusion of liquids and solids. Smallness and thermal fluctuations provide the basic reasons of the Brownian motion of nanoparticles. Increased importance of thermal fluctuations and configuration entropy at the nanoscale give rise to superelasticity, entropic elasticity (entropic forces), and other exotic types of elasticity of nanostructures. Aspects of configuration entropy are also of great interest in the context self-organization and cooperative behavior of open nanosystems.

Quantum effects determine *forces of interaction* between individual atoms in physical objects, which are introduced in nanomechanics by means of some averaged mathematical models called *interatomic potentials*.

Subsequent utilization of the interatomic potentials within the classical multibody dynamics provide deterministic mechanical models of nano structures and systems at the atomic scale/resolution. Numerical methods of solution of these models are called *molecular dynamics* (MD), and sometimes *molecular mechanics* (especially, in relation to statically equilibrated (still) models). Non-deterministic numerical approaches include Monte Carlo, Kinetic More-Carlo (KMC), and other methods. Contemporary numerical tools include also hybrid *multiscale approaches* allowing concurrent or sequential utilization of the atomistic scale methods (usually, MD) with the continuum (macro) scale methods (usually, field emission microscopy) within a single mathematical model. Development of these complex methods is a separate subject of applied mechanics research.

Quantum effects also determine novel electrical, optical and chemical properties of nanostructures, and therefore they find even greater attention in adjacent areas of nanoscience and nanotechnology, such as nanoelectronics, advanced energy systems, and nanobiotechnology.

# Molecular Scale Electronics

Molecular scale electronics, also called single-molecule electronics, is a branch of nanotechnology that uses single molecules, or nanoscale collections of single molecules, as electronic components. Because single molecules constitute the smallest stable structures imaginable this miniaturization is the ultimate goal for shrinking electrical circuits.

The field is often referred to as simply "molecular electronics", but this term is also used to refer to the distantly related field of conductive polymers and organic electronics, which uses the properties of molecules to affect the bulk properties of a material. A nomenclature distinction has been suggested so that *molecular materials for electronics* refers to this latter field of bulk applications, while *molecular scale electronics* refers to the nanoscale single-molecule applications discussed here.

## Fundamental Concepts

Conventional electronics have traditionally been made from bulk materials. Ever since their invention in 1958 the performance and complexity of integrated circuits has been growing exponentially (a trend also known as Moore's law) as feature sizes of the embedded components have shrink accordingly. As the structures become smaller, the sensitivity for deviations increases and in a few generations, when the minimum feature sizes reaches 13 nm, the composition of the devices must be controlled to a precision of a few atoms  for the devices to work. With the bulk approach becoming increasingly demanding and expensive as it nears its inherent limitats, the idea was born that the components could instead be built up atom for atom in a chemistry lab (bottom up) as opposed to carving them out of bulk material (top down). This is the idea behind molecular electronics, with the ultimate miniaturization being components contained in single molecules.

In single-molecule electronics, the bulk material is replaced by single molecules. That is, instead of creating structures by removing or applying material after a pattern scaffold, the atoms are put together in a chemistry lab. In this way billions of billions of copies are made simultaneously (typically more than $10^{20}$ molecules are made at once) while the composition of molecules are controlled down to the last atom. The molecules utilized have properties that resemble traditional electronic components such as a wire, transistor or rectifier.

Single-molecule electronics is an emerging field, and entire electronic circuits consisting exclusively of molecular sized compounds are still very far from being realized. However, the unceasing demand for more computing power together with the inherent limitations of the present day lithographic methods make the transition seem unavoidable. Currently, the focus is on discovering molecules with interesting properties and on finding ways to obtain reliable and reproducible contacts between the molecular components and the bulk material of the electrodes.

## Theoretical Basis

Molecular electronics operates in the quantum realm of distances less than 100 nanometers. The miniaturization down to single molecules brings the scale down to a regime where quantum effects are important. As opposed to the case in conventional electronic components, where electrons can be filled in or drawn out more or less like a continuous flow of charge, the transfer of a single electron alters the system significantly. This means that when an electron has been transferred from the source electrode to the molecule, the molecule gets charged up and makes it much harder for the next one to transfer. The significant amount of energy due to charging must be taken into account when making calculations about the electronic properties of the setup and is highly sensitive to distances to conducting surfaces nearby.

The theory of single-molecule devices is particularly interesting since the system under consideration is an open quantum system in nonequilibrium (driven by voltage). In the low bias voltage regime, the nonequilibrium nature of the molecular junction can be ignored, and the current-voltage characteristics of the device can be calculated using the equilibrium electronic structure of the system. However, in stronger bias regimes a more sophisticated treatment is required, as there is no longer a variational principle. In the elastic tunneling case (where the passing electron does not exchange energy with the system), the formalism of Rolf Landauer can be used to calculate the transmission through the system as a function of bias voltage, and hence the current. In inelastic tunneling, an elegant formalism based on the non-equilibrium Green's functions of Leo Kadanoff and Gordon Baym, and independently by Leonid Keldysh was put forth by Ned Wingreen and Yigal Meir. This Meir-Wingreen formulation has been used to great success in the molecular electronics community to examine the more difficult and interesting cases where the transient electron exchanges energy with the molecular system (for example through electron-phonon coupling or electronic excitations).

Further, connecting single molecules reliably to a larger scale circuit has proven a great challenge, and constitutes a significant hindrance to commercialization.

## Examples

Common for molecules utilized in molecular electronics is that the structures contain a lot of alternating double and single bonds. The reason for this is that such a pattern delocalizes the molecular orbitals making it possible for electrons to move freely over the conjugated area.

## Wires

The sole purpose of molecular wires is to electrically connect different parts of a molecular electrical circuit. As the assembly of these and their connection to a macroscopic circuit is still not mastered, the focus of research in single-molecule electronics is primarily on the functionalized molecules: molecular wires are characterized by containing no functional groups and are hence composed of plain repetitions of a conjugated building block. Among these are the carbon nanotubes that are quite large compared to the other suggestions but have shown very promising electrical properties.

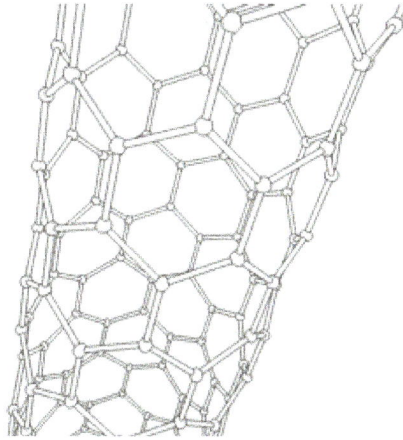

This animation of a rotating carbon nanotube shows its 3D structure.

The main problem with the molecular wires is to obtain good electrical contact with the electrodes so that the electrons can move freely in and out of the wire.

## Transistors

Single-molecule transistors are fundamentally different from the ones known from bulk electronics. The gate in a conventional (field-effect) transistor determines the conductance between the source and drain electrode by controlling the density of charge carriers between them, whereas the gate in a single-molecule transistor controls the feasibility of a single electron to jump on and off the molecule by modifying the energy of the molecular orbitals. One of the effects of this difference is that the single-molecule transistor is almost binary: it is either ON or OFF. This opposes its bulk counterparts, which have quadratic responses to gate voltage.

It is the quantization of charge into electrons that is responsible for the markedly different behavior compared to bulk electronics. Because of the size of a single molecule, the charging due to a single electron is significant and provides a mean to turn the transistor ON or OFF. For this to work, the electronic orbitals on the transistor molecule cannot be too well integrated with the orbitals on the electrodes. If they are, an electron cannot be said to be located on the molecule or the electrodes and the molecule will function as a wire.

A popular group of molecules, that can work as the semiconducting channel material in a molecular transistor, is the oligopolyphenylenevinylenes (OPVs) that works by the Coulomb blockade mechanism when placed between the source and drain electrode in an appropriate way. Fullerenes work by the same mechanism and have also been commonly utilized.

Semiconducting carbon nanotubes have also been demonstrated to work as channel material but although molecular, these molecules are sufficiently large to behave almost as bulk semiconductors.

The size of the molecules and the low temperature the measurements are being conducted at makes the quantum mechanical states well defined. It is therefore being researched if the quantum mechanical properties can be used for more advanced purposes than simple transistors (e.g. spintronics).

Physicists at the University of Arizona, in collaboration with chemists from the University of Madrid, have designed a single-molecule transistor using a ring-shaped molecule similar to benzene. Physicists at Canada's National Institute for Nanotechnology have designed a single-molecule transistor using styrene. Both groups expect (their designs have yet to be experimentally verified) their respective devices to function at room temperature, and to be controlled by a single electron.

## Rectifiers (Diodes)

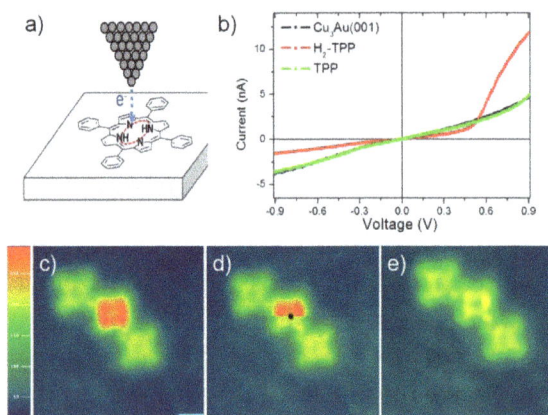

Hydrogen can be removed from individual $H_2TPP$ molecules by applying excess voltage to the tip of a scanning tunneling microscope (STAM, a); this removal alters the current-voltage (I-V) curves of TPP molecules, measured using the same STM tip, from diode-like (red curve in b) to resistor-like (green curve). Image (c) shows a row of TPP, $H_2TPP$ and TPP molecules. While scanning image (d), excess voltage was applied to $H_2TPP$ at the black dot, which instantly removed hydrogen, as shown in the bottom part of (d) and in the re-scan image (e). Such manipulations can be used in single-molecule electronics.

Molecular rectifiers are mimics of their bulk counterparts and have an asymmetric construction so that the molecule can accept electrons in one end but not the other. The molecules have an electron donor (D) in one end and an electron acceptor (A) in the other. This way, the unstable state $D^+ - A^-$ will be more readily made than $D^- - A^+$. The result is that an electric current can be drawn through the molecule if the electrons are added through the acceptor end, but not so easily if the reverse is attempted.

## Techniques

One of the biggest problems with measuring on single molecules is to establish reproducible electrical contact with only one molecule and doing so without shortcutting the electrodes. Because the current photolithographic technology is unable to produce electrode gaps small enough to contact both ends of the molecules tested (in the order of nanometers) alternative strategies are put into use.

## Molecular Gaps

One way to produce electrodes with a molecular sized gap between them is break junctions, in which a thin electrode is stretched until it breaks. Another is electromigration. Here a current is led through a thin wire until it melts and the atoms migrate to produce the gap. Further, the reach of conventional photolithography can be enhanced by chemically etching or depositing metal on the electrodes.

Probably the easiest way to conduct measurements on several molecules is to use the tip of a scanning tunneling microscope (STM) to contact molecules adhered at the other end to a metal substrate.

## Anchoring

A popular way to anchor molecules to the electrodes is to make use of sulfur's high affinity to gold. In these setups, the molecules are synthesized so that sulfur atoms are placed strategically to function as crocodile clips connecting the molecules to the gold electrodes. Though useful, the anchoring is non-specific and thus anchors the molecules randomly to all gold surfaces. Further, the contact resistance is highly dependent on the precise atomic geometry around the site of anchoring and thereby inherently compromises the reproducibility of the connection.

To circumvent the latter issue, experiments has shown that fullerenes could be a good candidate for use instead of sulfur because of the large conjugated $\pi$-system that can electrically contact many more atoms at once than a single atom of sulfur.

## Fullerene Nanoelectronics

In polymers, classical organic molecules are composed of both carbon and hydrogen (and sometimes additional compounds such as nitrogen, chlorine or sulphur). They are obtained from petrol and can often be synthesized in large amounts. Most of these molecules are insulating when their length exceeds a few nanometers. However, naturally occurring carbon is conducting. In particular, graphite (recovered from coal or encountered naturally) is conducting. From a theoretical point of view, graphite is a semi-metal, a category in between metals and semi-conductors. It has a layered structure, each sheet being one atom thick. Between each sheet, the interactions are weak enough to allow an easy manual cleavage.

Tailoring the graphite sheet to obtain well defined nanometer-sized objects remains a challenge. However, by the close of the twentieth century, chemists were exploring methods to fabricate extremely small graphitic objects that could be considered single molecules. After studying the interstellar conditions under which carbon is known to form clusters, Richard Smalley's group (Rice University, Texas) set up an experiment in which graphite was vaporized using laser irradiation. Mass spectrometry revealed that clusters containing specific "magic numbers" of atoms were stable, in particular those clusters of 60 atoms. Harry Kroto, an English chemist who assisted in the experiment, suggested a possible geometry for these clusters – atoms covalently bound with the exact symmetry of a soccer ball. Coined buckminsterfullerenes, buckyballs or $C_{60}$, the clusters retained some properties of graphite, such as conductivity. These objects were rapidly envisioned as possible building blocks for molecular electronics.

## Problems

### Artifacts

When trying to measure electronic characteristics of molecules, artificial phenomena can occur that can be hard to distinguish from truly molecular behavior. Before they were discovered these artifacts have mistakenly been published as being features pertaining to the molecules in question.

Applying a voltage drop in the order of volts across a nanometer sized junction results in a very strong electrical field. The field can cause metal atoms to migrate and eventually close the gap by a thin filament, which can be broken again when carrying a current. The two levels of conductance imitate molecular switching between a conductive and an isolating state of a molecule.

Another encountered artifact is when the electrodes undergo chemical reactions due to the high field strength in the gap. When the bias is reversed the reaction will cause hysteresis in the measurements that can be interpreted as being of molecular origin.

A metallic grain between the electrodes can act as a single electron transistor by the mechanism described above thus resembling the characteristics of a molecular transistor. This artifact is especially common with nanogaps produced by the electromigration technique.

### Commercialization

One of the biggest hindrances for single-molecule electronics to be commercially exploited is the lack of techniques to connect a molecular sized circuit to bulk electrodes in a way that gives reproducible results. At the current state, the difficulty of connecting single molecules vastly outweighs any possible performance increase that could be gained from such shrinkage. The picture becomes even worse if the molecules are to have a certain spatial orientation and/or have multiple poles to connect.

Also problematic is the fact that some measurements on single molecules are carried out in cryogenic temperatures (close to absolute zero), which is very energy consuming. This is done to reduce signal noise enough to measure the faint currents of single molecules.

## History and Recent Progress

In their discussion of so-called "donor-acceptor" complexes in the 1940s, Robert Mulliken and Albert Szent-Gyorgi advanced the concept of charge transfer in molecules. They subsequently further refined the study of both charge transfer and energy transfer in molecules. Likewise, a 1974 paper from Mark Ratner and Ari Aviram illustrated a theoretical molecular rectifier. In 1988, Aviram described in detail a theoretical single-molecule field-effect transistor. Further concepts were proposed by Forrest Carter of the Naval Research Laboratory, including single-molecule logic gates. A wide range of ideas were presented, under his aegis, at a conference entitled *Molecular Electronic Devices* in 1988. These were all theoretical constructs and not concrete devices. The *direct* measurement of the electronic characteristics of individual molecules awaited the development of methods for making molecular-scale electrical contacts. This was no easy task. Thus, the first experiment directly-measuring the conductance of a single molecule was only reported in 1995 on a single $C_{60}$ molecule by C. Joachim and J. K. Gimzewsky in their seminal Physical Revie Letter paper and later in 1997 by Mark Reed and co-workers on a few hundred molecules. Since then, this branch of the field has progressed rapidly. Likewise, as it has become possible to measure such properties directly, the theoretical predictions of the early workers have been substantially confirmed.

Recent progress in nanotechnology and nanoscience has facilitated both experimental and theoretical study of molecular electronics. In particular, the development of the scanning tunneling microscope (STM) and later the atomic force microscope (AFM) have facilitated manipulation of single-molecule electronics. In addition, theoretical advances in molecular electronics have facilitated further understanding of non-adiabatic charge transfer events at electrode-electrolyte interfaces.

The concept of molecular electronics was first published in 1974 when Aviram and Ratner suggested an organic molecule that could work as a rectifier. Having both huge commercial and fundamental interest much effort was put into proving its feasibility and 16 years later in 1990 the first demonstration of an intrinsic molecular rectifier was realized by Ashwell and coworkers for a thin film of molecules.

The first measurement of the conductance of a single molecule was realised in 1994 by C. Joachim and J. K. Gimzewski and published in 1995. This was the conclusion of 10 years of research started at IBM TJ Watson, using the scanning tunnelling micro scope tip apex to switch a single molecule as already explored by A. Aviram, C. Joachim and M. Pomerantz at the end of the 80's. The trick was to use an UHV Scanning Tunneling microscope to allow the tip apex to gently touch the top of a single C

60 molecule adsorbed on an Au(110) surface. A resistance of 55 MOhms was recorded together with a low voltage linear I-V. The contact was certified by recording the I-z current distance characteristic, which allows the measurement of the deformation of the C 60 cage under contact. This first experiment was followed by the reported result using a mechanical break junction approach to connect two gold electrodes to a sulfur-terminated molecular wire by Mark Reed and James Tour in 1997.

A single-molecule amplifier was implemented by C. Joachim and J.K. Gimzewski in IBM Zurich. This experiment involving a single C 60 molecule demonstrated that a single C 60 molecule can provide gain in a circuit just by playing with through C 60 intramolecular quantum interference effects.

A collaboration of researchers at HP and UCLA, led by James Heath, Fraser Stoddart, R. Stanley Williams, and Philip Kuekes, has developed molecular electronics based on rotaxanes and catenanes.

Work is also being done on the use of single-wall carbon nanotubes as field-effect transistors. Most of this work is being done by IBM.

Some specific reports of a field-effect transistor based on molecular self-assembled monolayers were shown to be fraudulent in 2002 as part of the Schön scandal.

Until recently entirely theoretical, the Aviram-Ratner model for a unimolecular rectifier has been unambiguously-confirmed in experiments by a group led by Geoffrey J. Ashwell at Bangor University, UK. Many rectifying molecules have so far been identified, and the number and efficiency of these systems is expanding rapidly.

Supramolecular electronics is a new field that tackles electronics at a supramolecular level.

An important issue in molecular electronics is the determination of the resistance of a single molecule (both theoretical and experimental). For example, Bumm, et al. used STM to analyze a single molecular switch in a self-assembled monolayer to determine how conductive such a molecule can be. Another problem faced by this field is the difficulty of performing direct characterization since imaging at the molecular scale is often difficult in many experimental devices.

## Impact of Nanotechnology

The impact of nanotechnology extends from its medical, ethical, mental, legal and environmental applications, to fields such as engineering, biology, chemistry, computing, materials science, and communications.

Major benefits of nanotechnology include improved manufacturing methods, water purification systems, energy systems, physical enhancement, nanomedicine, better food production methods, nutrition and large-scale infrastructure auto-fabrication.

Nanotechnology's reduced size may allow for automation of tasks which were previously inaccessible due to physical restrictions, which in turn may reduce labor, land, or maintenance requirements placed on humans.

Potential risks include environmental, health, and safety issues; transitional effects such as displacement of traditional industries as the products of nanotechnology become dominant, which are of concern to privacy rights advocates. These may be particularly important if potential negative effects of nanoparticles are overlooked.

Whether nanotechnology merits special government regulation is a controversial issue. Regulatory bodies such as the United States Environmental Protection Agency and the Health and Consumer Protection Directorate of the European Commission have started dealing with the potential risks of nanoparticles. The organic food sector has been the first to act with the regulated exclusion of engineered nanoparticles from certified organic produce, firstly in Australia and the UK, and more recently in Canada, as well as for all food certified to Demeter International standards.

## Overview

The presence of nanomaterials (materials that contain nanoparticles) is not in itself a threat. It is only certain aspects that can make them risky, in particular their mobility and their increased reactivity. Only if certain properties of certain nanoparticles were harmful to living beings or the environment would we be faced with a genuine hazard. In this case it can be called nanopollution.

In addressing the health and environmental impact of nanomaterials we need to differentiate between two types of nanostructures: (1) Nanocomposites, nanostructured surfaces and nanocomponents (electronic, optical, sensors etc.), where nanoscale particles are incorporated into a substance, material or device ("fixed" nano-particles); and (2) "free" nanoparticles, where at some stage in production or use individual nanoparticles of a substance are present. These free nanoparticles could be nanoscale species of elements, or simple compounds, but also complex compounds where for instance a nanoparticle of a particular element is coated with another substance ("coated" nanoparticle or "core-shell" nanoparticle).

There seems to be consensus that, although one should be aware of materials containing fixed nanoparticles, the immediate concern is with free nanoparticles.

Nanoparticles are very different from their everyday counterparts, so their adverse effects cannot be derived from the known toxicity of the macro-sized material. This poses significant issues for addressing the health and environmental impact of free nanoparticles.

To complicate things further, in talking about nanoparticles it is important that a powder or liquid containing nanoparticles almost never be monodisperse, but contain instead a range of particle sizes. This complicates the experimental analysis as larger nanoparticles might have different properties from smaller ones. Also, nanoparticles show a tendency to aggregate, and such aggregates often behave differently from individual nanoparticles.

## Health Impact

The health impacts of nanotechnology are the possible effects that the use of nanotechnological materials and devices will have on human health. As nanotechnology is an emerging field, there is great debate regarding to what extent nanotechnology will benefit or pose risks for human health. Nanotechnology's health impacts can be split into two aspects: the potential for nanotechnological innovations to have medical applications to cure disease, and the potential health hazards posed by exposure to nanomaterials.

## Medical Applications

Nanomedicine is the medical application of nanotechnology. The approaches to nanomedicine range from the medical use of nanomaterials, to nanoelectronic biosensors, and even possible future applications of molecular nanotechnology. Nanomedicine seeks to deliver a valuable set of research tools and clinically helpful devices in the near future. The National Nanotechnology Initiative expects new commercial applications in the pharmaceutical industry that may include advanced drug delivery systems, new therapies, and in vivo imaging. Neuro-electronic interfaces and other nanoelectronics-based sensors are another active goal of research. Further down the line, the speculative field of molecular nanotechnology believes that cell repair machines could revolutionize medicine and the medical field.

Nanomedicine seeks to deliver a set of research tools and clinical devices in the near future. The National Nanotechnology Initiative expects new commercial applications in the pharmaceutical industry that may include advanced drug delivery systems, new therapies, and *in vivo* imaging. Neuro-electronic interfaces and other nanoelectronics-based sensors are another active goal of research. Further down the line, the speculative field of molecular nanotechnology believes that cell repair machines could revolutionize medicine and the medical field.

Nanomedicine research is directly funded, with the US National Institutes of Health in 2005 funding a five-year plan to set up four nanomedicine centers. In April 2006, the journal Nature Materials estimated that 130 nanotech-based drugs and delivery systems were being developed worldwide. Nanomedicine is a large industry, with nanomedicine sales reaching $6.8 billion in 2004. With over 200 companies and 38 products worldwide, a minimum of $3.8 billion in nanotechnology R&D is being invested every year. As the nanomedicine industry continues to grow, it is expected to have a significant impact on the economy.

# Health Hazards

Nanotoxicology is the field which studies potential health risks of nanomaterials. The extremely small size of nanomaterials means that they are much more readily taken up by the human body than larger sized particles. How these nanoparticles behave inside the organism is one of the significant issues that needs to be resolved. The behavior of nanoparticles is a function of their size, shape and surface reactivity with the surrounding tissue. Apart from what happens if non-degradable or slowly degradable nanoparticles accumulate in organs, another concern is their potential interaction with biological processes inside the body: because of their large surface, nanoparticles on exposure to tissue and fluids will immediately adsorb onto their surface some of the macromolecules they encounter. The large number of variables influencing toxicity means that it is difficult to generalise about health risks associated with exposure to nanomaterials – each new nanomaterial must be assessed individually and all material properties must be taken into account. Health and environmental issues combine in the workplace of companies engaged in producing or using nanomaterials and in the laboratories engaged in nanoscience and nanotechnology research. It is safe to say that current workplace exposure standards for dusts cannot be applied directly to nanoparticle dusts.

The extremely small size of nanomaterials also means that they are much more readily taken up by the human body than larger sized particles. How these nanoparticles behave inside the body is one of the issues that needs to be resolved. The behavior of nanoparticles is a function of their size, shape and surface reactivity with the surrounding tissue. They could cause overload on phagocytes, cells that ingest and destroy foreign matter, thereby triggering stress reactions that lead to inflammation and weaken the body's defense against other pathogens. Apart from what happens if non-degradable or slowly degradable nanoparticles accumulate in organs, another concern is their potential interaction with biological processes inside the body: because of their large surface, nanoparticles on exposure to tissue and fluids will immediately adsorb onto their surface some of the macromolecules they encounter. This may, for instance, affect the regulatory mechanisms of enzymes and other proteins.

The National Institute for Occupational Safety and Health has conducted initial research on how nanoparticles interact with the body's systems and how workers might be exposed to nano-sized particles in the manufacturing or industrial use of nanomaterials. NIOSH currently offers interim guidelines for working with nanomaterials consistent with the best scientific knowledge. At The National Personal Protective Technology Laboratory of NIOSH, studies investigating the filter penetration of nanoparticles on NIOSH-certified and EU marked respirators, as well as non-certified dust masks have been conducted. These studies found that the most penetrating particle size range was between 30 and 100 nanometers, and leak size was the largest factor in the number of nanoparticles found inside the respirators of the test dummies.

Other properties of nanomaterials that influence toxicity include: chemical composition, shape, surface structure, surface charge, aggregation and solubility, and the presence or absence of functional groups of other chemicals. The large number of variables influencing toxicity means that it is difficult to generalise about health risks associated with exposure to nanomaterials – each new nanomaterial must be assessed individually and all material properties must be taken into account.

Literature reviews have been showing that release of engineered nanoparticles and incurred personal exposure can happen during different work activities. The situation alerts regulatory bodies to necessitate prevention strategies and regulations at nanotechnology workplaces.

## Environmental Impact

The environmental impact of nanotechnology is the possible effects that the use of nanotechnological materials and devices will have on the environment. As nanotechnology is an emerging field, there is great debate regarding to what extent industrial and commercial use of nanomaterials will affect organisms and ecosystems.

Nanotechnology's environmental impact can be split into two aspects: the potential for nanotechnological innovations to help improve the environment, and the possibly novel type of pollution that nanotechnological materials might cause if released into the environment.

## Environmental Applications

Green nanotechnology refers to the use of nanotechnology to enhance the environmental sustainability of processes producing negative externalities. It also refers to the use of the products of nanotechnology to enhance sustainability. It includes making green nano-products and using nano-products in support of sustainability. Green nanotechnology has been described as the development of clean technologies, "to minimize potential environmental and human health risks associated with the manufacture and use of nanotechnology products, and to encourage replacement of existing products with new nano-products that are more environmentally friendly throughout their lifecycle."

Green nanotechnology has two goals: producing nanomaterials and products without harming the environment or human health, and producing nano-products that provide solutions to environmental problems. It uses existing principles of green chemistry and green engineering to make nanomaterials and nano-products without toxic ingredients, at low temperatures using less energy and renewable inputs wherever possible, and using lifecycle thinking in all design and engineering stages.

## Pollution

Nanopollution is a generic name for all waste generated by nanodevices or during the nanomaterials manufacturing process. Nanowaste is mainly the group of particles that are released into the environment, or the particles that are thrown away when still on their products.

## Societal Impact

Beyond the toxicity risks to human health and the environment which are associated with first-generation nanomaterials, nanotechnology has broader societal impact and poses broader social challenges. Social scientists have suggested that nanotechnology's social issues should be understood and assessed not simply as "downstream" risks or impacts. Rather, the challenges should be factored into "upstream" research and decision-making in order to ensure technology development that meets social objectives.

Many social scientists and organizations in civil society suggest that technology assessment and governance should also involve public participation.

Over 800 nano-related patents were granted in 2003, with numbers increasing to nearly 19,000 internationally by 2012. Corporations are already taking out broad-ranging patents on nanoscale discoveries and inventions. For example, two corporations, NEC and IBM, hold the basic patents on carbon nanotubes, one of the current cornerstones of nanotechnology. Carbon nanotubes have a wide range of uses, and look set to become crucial to several industries from electronics and computers, to strengthened materials to drug delivery and diagnostics. Carbon nanotubes are poised to become a major traded commodity with the potential to replace major conventional raw materials.

Nanotechnologies may provide new solutions for the millions of people in developing countries who lack access to basic services, such as safe water, reliable energy, health care, and education. The 2004 UN Task Force on Science, Technology and Innovation noted that some of the advantages of nanotechnology include production using little labor, land, or maintenance, high productivity, low cost, and modest requirements for materials and energy. However, concerns are frequently raised that the claimed benefits of nanotechnology will not be evenly distributed, and that any benefits (including technical and/or economic) associated with nanotechnology will only reach affluent nations.

Longer-term concerns center on the impact that new technologies will have for society at large, and whether these could possibly lead to either a post-scarcity economy, or alternatively exacerbate the wealth gap between developed and developing nations. The effects of nanotechnology on the society as a whole, on human health and the environment, on trade, on security, on food systems and even on the definition of "human", have not been characterized or politicized.

## Regulation

Significant debate exists relating to the question of whether nanotechnology or nano-technology-based products merit special government regulation. This debate is related to the circumstances in which it is necessary and appropriate to assess new substances prior to their release into the market, community and environment.

Regulatory bodies such as the United States Environmental Protection Agency and the Food and Drug Administration in the U.S. or the Health & Consumer Protection Directorate of the European Commission have started dealing with the potential risks posed by nanoparticles. So far, neither engineered nanoparticles nor the products and materials that contain them are subject to any special regulation regarding production, handling or labelling. The Material Safety Data Sheet that must be issued for some materials often does not differentiate between bulk and nanoscale size of the material in question and even when it does these MSDS are advisory only.

Limited nanotechnology labeling and regulation may exacerbate potential human and environmental health and safety issues associated with nanotechnology. It has been argued that the development of comprehensive regulation of nanotechnology will be vital to ensure that the potential risks associated with the research and commercial application of nanotechnology do not overshadow its potential benefits. Regulation may also be required to meet community expectations about responsible development of nanotechnology, as well as ensuring that public interests are included in shaping the development of nanotechnology.

In "The Consumer Product Safety Commission and Nanotechnology," E. Marla Felcher suggests that the Consumer Product Safety Commission, which is charged with protecting the public against unreasonable risks of injury or death associated with consumer products, is ill-equipped to oversee the safety of complex, high-tech products made using nanotechnology.

## Regulation of Nanotechnology

Because of the ongoing controversy on the implications of nanotechnology, there is significant debate concerning whether nanotechnology or nanotechnology-based products merit special government regulation. This mainly relates to when to assess new substances prior to their release into the market, community and environment.

Nanotechnology refers to an increasing number of commercially available products – from socks and trousers to tennis racquets and cleaning cloths. Such nanotechnologies and their accompanying industries have triggered calls for increased community participation and effective regulatory arrangements. However, these calls have presently not led to such comprehensive regulation to oversee research and the commercial application of nanotechnologies, or any comprehensive labeling for products that contain nanoparticles or are derived from nano-processes.

Regulatory bodies such as the United States Environmental Protection Agency and the Food and Drug Administration in the U.S. or the Health and Consumer Protection Directorate of the European Commission have started dealing with the potential risks posed by nanoparticles. So far, neither engineered nanoparticles nor the products and materials that contain them are subject to any special regulation regarding production, handling or labelling.

## Managing Risks: Human and Environmental Health and Safety

Studies of the health impact of airborne particles generally shown that for toxic materials, smaller particles are more toxic. This is due in part to the fact that, given the same mass per volume, the dose in terms of particle numbers increases as particle size decreases.

Based upon available data, it has been argued that current risk assessment methodologies are not suited to the hazards associated with nanoparticles; in particular, existing toxicological and eco-toxicological methods are not up to the task; exposure evaluation (dose) needs to be expressed as quantity of nanoparticles and/or surface area rather than simply mass; equipment for routine detecting and measuring nanoparticles in air, water, or soil is inadequate; and very little is known about the physiological responses to nanoparticles.

Regulatory bodies in the U.S. as well as in the EU have concluded that nanoparticles form the potential for an entirely new risk and that it is necessary to carry out an extensive analysis of the risk. The challenge for regulators is whether a matrix can be developed which would identify nanoparticles and more complex nanoformulations which are likely to have special toxicological properties or whether it is more reasonable for each particle or formulation to be tested separately.

The International Council on Nanotechnology maintains a database and Virtual Journal of scientific papers on environmental, health and safety research on nanoparticles. The database currently has over 2000 entries indexed by particle type, exposure pathway and other criteria. The Project on Emerging Nanotechnologies (PEN) currently lists 807 products that manufacturers have voluntarily identified that use nanotechnology. No labeling is required by the FDA so that number could be significantly higher. "The use of nanotechnology in consumer products and industrial applications is growing rapidly, with the products listed in the PEN inventory showing just the tip of the iceberg" according to PEN Project Director David Rejeski . A list of those products that have been voluntarily disclosed by their manufacturers is located here .

The Material Safety Data Sheet that must be issued for certain materials often does not differentiate between bulk and nanoscale size of the material in question and even when it does these MSDS are advisory only.

## Democratic Governance

Many argue that government has a responsibility to provide opportunities for the public to be involved in the development of new forms of science and technology. Community engagement can be achieved through various means or mechanisms. An online journal article identifies traditional approaches such as referenda, consultation documents, and advisory committees that include community members and other stakeholders. Other conventional approaches include public meetings and "closed" dialog with stakeholders. More contemporary engagement processes that have been employed to include community members in decisions about nanotechnology include citizens' juries and consensus conferences. Leach and Scoones (2006, p. 45) argue that since that "most debates about science and technology options involve uncertainty, and often ignorance, public debate about regulatory regimes is essential."

It has been argued that limited nanotechnology labeling and regulation may exacerbate potential human and environmental health and safety issues associated with nanotechnology, and that the development of comprehensive regulation of nanotechnology will be vital to ensure that the potential risks associated with the research and commercial application of nanotechnology do not overshadow its potential benefits. Regulation may also be required to meet community expectations about responsible development of nanotechnology, as well as ensuring that public interests are included in shaping the development of nanotechnology.

Community education, engagement and consultation tend to occur "downstream": once there is at least a moderate level of awareness, and often during the process of disseminating and adapting technologies. "Upstream" engagement, by contrast, occurs much earlier in the innovation cycle and involves: "dialogue and debate about future technology options and pathways, bringing the often expert-led approaches to horizon scanning, technology foresight and scenario planning to involve a wider range of perspectives and inputs." Daniel Sarewitz Director of Arizona State University's Consortium on Science, Policy and Outcomes, argues that "by the time new devices reach the stage of commercialization and regulation, it is usually too late to alter them to correct problems." However, Xenos, et al. argue that upstream engagement can be utilized in this area through anticipated discussion with peers. Upstream engagement in this sense is "meant to create the best possible conditions for sound policy making and public judgments based on carefull assessment of objective information". Discussion may act as a catalyst for upstream engagement by prompting accountability for individuals to seek and process additional information ("anticipatory elaboration"). However, though anticipated discussion did lead to participants seeking further information, Xenos et al. found that factual information was not primarily sought out; instead, individuals sought out opinion pieces and editorials.

The stance that the research, development and use of nanotechnology should be subject to control by the public sector is sometimes referred to as nanosocialism.

## Newness

The question of whether nanotechnology represents something 'new' must be answered to decide how best nanotechnology should be regulated. The Royal Society recommended that the UK government assess chemicals in the form of nanoparticles or nanotubes as new substances. Subsequent to this, in 2007 a coalition of over forty groups called for nanomaterials to be classified as new substances, and regulated as such.

Despite these recommendations, chemicals comprising nanoparticles that have previously been subject to assessment and regulation may be exempt from regulation, regardless of the potential for different risks and impacts. In contrast, nanomaterials are often recognized as 'new' from the perspective of intellectual property rights (IPRs), and as such are commercially protected via patenting laws.

There is significant debate about who is responsible for the regulation of nanotechnology. While some non-nanotechnology specific regulatory agencies currently cover some products and processes (to varying degrees) – by "bolting on" nanotechnology to existing regulations – there are clear gaps in these regimes. This enables some nanotechnology applications to figuratively "slip through the cracks" without being covered by any regulations. An example of this has occurred in the US, and involves nanoparticles of titanium dioxide (TIo2) for use in sunscreen where they create a clearer cosmetic appearance. In this case, the US Food and Drug Administration (FDA) reviewed the immediate health effects of exposure to nanoparticles of titanium dioxide (TIo2) for consumers. However, they did not review its impacts for aquatic ecosystems when the sunscreen rubs off, nor did the EPA, or any other agency. Similarly the Australian equivalent of the FDA, the Therapeutic Goods Administration (TGA) approved the use of nanoparticles in sunscreens (without the requirement for package labelling) after a thorough review of the literature, on the basis that although nanoparticles of TIo2 and zinc oxide (ZNo) in sunscreens do produce free radicals and oxidative DNA damage *in vitro*, such particles were unlikely to pass the dead outer cells of the stratum corneum of human skin; a finding which some academics have argued seemed not to apply the precautionary principle in relation to prolonged use on children with cut skin, the elderly with thin skin, people with diseased skin or use over flexural creases. Doubts over the TGA's decision were raised with publication of a paper showing that the uncoated anatase form of TIo2 used in some Australian sunscreens caused a photocatalytic reaction that degraded the surface of newly installed prepainted steel roofs in places where they came in contact with the sunscreen coated hands of workmen. Such gaps in regulation are likely to continue alongside the development and commercialization of increasingly complex second and third generation nanotechnologies.

Nanomedicines are just beginning to enter drug regulatory processes, but within a few decades could comprise a dominant group within the class of innovative pharmaceuticals, the current thinking of government safety and cost-effectiveness regulators appearing to be that these products give rise to few if any nano-specific issues. Some academics (such as Thomas Alured Faunce) have challenged that proposition and suggest that nanomedicines may create unique or heightened policy challenges for government systems of cost-effectiveness as well as safety regulation. There are also significant public good aspects to the regulation of nanotechnology, particularly with regard to ensuring that industry involvement in standard-setting does not become a means of reducing competition and that nanotechnology policy and regulation encourages new models of safe drug discovery and development more systematically targeted at the global burden of disease.

Self-regulation attempts may well fail, due to the inherent conflict of interest in asking any organization to police itself. If the public becomes aware of this failure, an external, independent organization is often given the duty of policing them, sometimes with highly punitive measures taken against the organization. The Food and Drug Administration notes that it only regulates on the basis of voluntary claims made by the product manufacturer. If no claims are made by a manufacturer, then the FDA may be unaware of nanotechnology being employed.

Yet regulations worldwide still fail to distinguish between materials in their nanoscale and bulk form. This means that nanomaterials remain effectively unregulated; there is no regulatory requirement for nanomaterials to face new health and safety testing or environmental impact assessment prior to their use in commercial products, if these materials have already been approved in bulk form. The health risks of nanomaterials are of particular concern for workers who may face occupational exposure to nanomaterials at higher levels, and on a more routine basis, than the general public.

## International Law

There is no international regulation of nanoproducts or the underlying nanotechnology. Nor are there any internationally agreed definitions or terminology for nanotechnology, no internationally agreed protocols for toxicity testing of nanoparticles, and no standardized protocols for evaluating the environmental impacts of nanoparticles.

Since products that are produced using nanotechnologies will likely enter international trade, it is argued that it will be necessary to harmonize nanotechnology standards across national borders. There is concern that some countries, most notably developing countries, will be excluded from international standards negotiations. The Institute for Food and Agricultural Standards notes that "developing countries should have a say in international nanotechnology standards development, even if they lack capacity to enforce the standards".

Concerns about monopolies and concentrated control and ownership of new nanotechnologies were raised in community workshops in Australia in 2004.

## Arguments Against Regulation

Wide use of the term nanotechnology in recent years has created the impression that regulatory frameworks are suddenly having to contend with entirely new challenges that they are unequipped to deal with. Many regulatory systems around the world already assess new substances or products for safety on a case by case basis, before they are permitted on the market. These regulatory systems have been assessing the safety of nanometre scale molecular arrangements for many years and many substances comprising nanometre scale particles have been in use for decades e.g. Carbon black, Titanium dioxide, Zinc oxide, Bentonite, Aluminum silicate, Iron oxides, Silicon dioxide, Diatomaceous earth, Kaolin, Talc, Montmorillonite, Magnesium oxide, Copper sulphate.

These existing approval frameworks almost universally use the best available science to assess safety and do not approve substances or products with an unacceptable risk benefit profile. One proposal is to simply treat particle size as one of the several parameters defining a substance to be approved, rather than creating special rules for all particles of a given size regardless of type. A major argument against special regulation of nanotechnology is that the projected applications with the greatest impact are far in the future, and it is unclear how to regulate technologies whose feasibility is speculative at this point. In the meantime, it has been argued that the immediate applications of nanomaterials raise challenges not much different from those of introducing any other new material, and can be dealt with by minor tweaks to existing regulatory schemes rather than sweeping regulation of entire scientific fields.

A truly precautionary approach to regulation could severely impede development in the field of nanotechnology safety studies are required for each and every nanoscience application. While the outcome of these studies can form the basis for government and international regulations, a more reasonable approach might be development of a risk matrix that identifies likely culprits.

## Response From Governments

### United Kingdom

In its seminal 2004 report *Nanoscience and Nanotechnologies: Opportunities and Uncertainties*, the United Kingdom's Royal Society concluded that:

> Many nanotechnologies pose no new risks to health and almost all the concerns relate to the potential impacts of deliberately manufactured nanoparticles and nanotubes that are free rather than fixed to or within a material... We expect the likelihood of nanoparticles or nanotubes being released from products in which they have been fixed or embedded (such as composites) to be low but have recommended that manufacturers assess this potential exposure risk for the lifecycle of the product and make their findings available to the relevant regulatory bodies... It is very unlikely

*that new manufactured nanoparticles could be introduced into humans in doses sufficient to cause the health effects that have been associated with [normal air pollution].*

but have recommended that nanomaterials be regulated as new chemicals, that research laboratories and factories treat nanomaterials "as if they were hazardous", that release of nanomaterials into the environment be avoided as far as possible, and that products containing nanomaterials be subject to new safety testing requirements prior to their commercial release.

The 2004 report by the UK Royal Society and Royal Academy of Engineers noted that existing UK regulations did not require additional testing when existing substances were produced in nanoparticulate form. The Royal Society recommended that such regulations were revised so that "chemicals produced in the form of nanoparticles and nanotubes be treated as new chemicals under these regulatory frameworks" (p.xi). They also recommended that existing regulation be modified on a precautionary basis because they expect that "the toxicity of chemicals in the form of free nanoparticles and nanotubes cannot be predicted from their toxicity in a larger form and... in some cases they will be more toxic than the same mass of the same chemical in larger form."

The Better Regulation Commission's earlier 2003 report had recommended that the UK Government:

1.  enable, through an informed debate, the public to consider the risks for themselves, and help them to make their own decisions by providing suitable information;

2.  be open about how it makes decisions, and acknowledge where there are uncertainties;

3.  communicate with, and involve as far as possible, the public in the decision making process;

4.  ensure it develops two-way communication channels; and

5.  take a strong lead over the handling of any risk issues, particularly information provision and policy implementation.

These recommendations were accepted in principle by the UK Government. Noting that there was "no obvious focus for an informed public debate of the type suggested by the Task Force", the UK government's response was to accept the recommendations.

The Royal Society's 2004 report identified two distinct governance issues:

1.  the "role and behaviour of institutions" and their ability to "minimise unintended consequences" through adequate regulation and

2. the extent to which the public can trust and play a role in determining the trajectories that nanotechnologies may follow as they develop.

## United States

Rather than adopt a new nano-specific regulatory framework, the United States' Food and Drug Administration (FDA) convenes an 'interest group' each quarter with representatives of FDA centers that have responsibility for assessment and regulation of different substances and products. This interest group ensures coordination and communication. A September 2009 FDA document called for identifying sources of nanomaterials, how they move in the environment, the problems they might cause for people, animals and plants, and how these problems could be avoided or mitigated.

The Bush administration in 2007 decided that no special regulations or labeling of nanoparticles were required. Critics derided this as treating consumers like a "guinea pig" without sufficient notice due to lack of labelling.

Berkeley, CA is currently the only city in the United States to regulate nanotechnology. Cambridge, MA in 2008 considered enacting a similar law, but the committee it instituted to study the issue Cambridge recommended against regulation in its final report, recommending instead other steps to facilitate information-gathering about potential effects of nanomaterials.

On December 10, 2008 the U.S. National Research Council released a report calling for more regulation of nanotechnology.

## California

Assembly Bill (AB) 289 (2006) authorizes the Department of Toxic Substances Control (DTSC) within the California Environmental Protection Agency and other agencies to request information on environmental and health impacts from chemical manufacturers and importers, including testing techniques.

In October 2008, the Department of Toxic Substances Control (DTSC), within the California Environmental Protection Agency, announced its intent to request information regarding analytical test methods, fate and transport in the environment, and other relevant information from manufacturers of carbon nanotubes. DTSC is exercising its authority under the California Health and Safety Code, Chapter 699, sections 57018-57020. These sections were added as a result of the adoption of Assembly Bill AB 289 (2006). They are intended to make information on the fate and transport, detection and analysis, and other information on chemicals more available. The law places the responsibility to provide this information to the Department on those who manufacture or import the chemicals.

On January 22, 2009, a formal information request letter was sent to manufacturers who produce or import carbon nanotubes in California, or who may export carbon nanotubes into the State. This letter constitutes the first formal implementation of the authorities placed into statute by AB 289 and is directed to manufacturers of carbon nanotubes, both industry and academia within the State, and to manufacturers outside California who export carbon nanotubes to California. This request for information must be met by the manufacturers within one year. DTSC is waiting for the upcoming January 22, 2010 deadline for responses to the data call-in.

The California Nano Industry Network and DTSC hosted a full-day symposium on November 16, 2009 in Sacramento, CA. This symposium provided an opportunity to hear from nanotechnology industry experts and discuss future regulatory considerations in California.

DTSC is expanding the Specific Chemical Information Call-in to members of the nano-metal oxides. Interested individuals are encouraged to visit their website for the latest up-to-date information.

On December 21, 2010, the Department of Toxic Substances Control (DTSC) initiated the second Chemical Information Call-in for six nanomaterials: nano cerium oxide, nano silver, nano titanium dioxide, nano zero valent iron, nano zinc oxide, and quantum dots. DTSC sent a formal information request letter to forty manufacturers who produce or import the six nanomaterials in California, or who may export them into the State. The Chemical Information Call-in is meant to identify information gaps of these six nanomaterials and to develop further knowledge of their analytical test methods, fate and transport in the environment, and other relevant information under California Health and Safety Code, Chapter 699, sections 57018-57020. DTSC completed the carbon nanotube information call-in in June 2010.

DTSC partners with University of California, Los Angeles (UCLA), Santa Barbara (UCSB), and Riverside (UCR), University of Southern California (USC), Stanford University, Center for Environmental Implications of Nanotechnology (CEIN), and The National Institute for Occupational Safety and Health (NIOSH) on safe nanomaterial handling practices.

DTSC is interested in expanding the Chemical Information Call-in to members of the bominated flame retardants, members of the methyl siloxanes, ocean plastics, nano-clay, and other emerging chemicals.

## European Union

The European Union has formed a group to study the implications of nanotechnology called the Scientific Committee on Emerging and Newly Identified Health Risks which has published a list of risks associated with nanoparticles.

Consequently, manufacturers and importers of carbon products, including carbon na-no-tubes will have to submit full health and safety data within a year or so in order to comply with REACH.

## Response from Advocacy Groups

In January 2008, a coalition of over 40 civil society groups endorsed a statement of principles calling for precautionary action related to nanotechnology. The coalition called for strong, comprehensive oversight of the new technology and its products in the International Center for Technology Assessment's report *Principles for the Oversight of Nanotechnologies and Nano materials*, which states:

> *Hundreds of consumer products incorporating nano-materials are now on the market, including cosmetics, sunscreens, sporting goods, clothing, electronics, baby and infant products, and food and food packaging. But evidence indicates that current nano-materials may pose significant health, safety, and environmental hazards. In addition, the profound social, economic, and ethical challenges posed by nano-scale technologies have yet to be addressed ... 'Since there is currently no government oversight and no labeling requirements for nano-products anywhere in the world, no one knows when they are exposed to potential nano-tech risks and no one is monitoring for potential health or environmental harm. That's why we believe oversight action based on our principles is urgent' ... This industrial boom is creating a growing nano-workforce which is predicted to reach two million globally by 2015. 'Even though potential health hazards stemming from exposure have been clearly identified, there are no mandatory workplace measures that require exposures to be assessed, workers to be trained, or control measures to be implemented,' explained Bill Kojola of the AFL-CIO. 'This technology should not be rushed to market until these failings are corrected and workers assured of their safety'"* also .

The group has urged action based on eight principles. They are 1) A Precautionary Foundation 2) Mandatory Nano-specific Regulations 3) Health and Safety of the Public and Workers 4) Environmental Protection 5) Transparency 6) Public Participation 7) Inclusion of Broader Impacts and 8) Manufacturer Liability.

Some NGOs, including Friends of the Earth, are calling for the formation of a separate nanotechnology specific regulatory framework for the regulation of nanotechnology. In Australia, Friends of the Earth propose the establishment of a Nanotechnology Regulatory Coordination Agency, overseen by a Foresight and Technology Assessment Board. The advantage of this arrangement is that it could ensure a centralized body of experts that are able to provide oversight across the range of nano-products and sectors. It is also argued that a centralized regulatory approach would simplify the regulatory environment, thereby supporting industry innovation. A National Nanotechnology Regulator could coordinate existing regulations related to nano-technology (including intellectual property, civil liberties, product safety, occupation health and safety, environmental and international law). Regulatory mechanisms could vary from "hard law at one extreme through licensing and codes of practice to 'soft' self-regulation and negotiation in order to influence behavior." The formation of national nanotechnology regulatory bodies may also assist in establishing global regulatory frameworks.

In early 2008, The UK's largest organic certifier, the Soil Association, announced that its organic standard would exclude nanotechnology, recognizing the associated human and environmental health and safety risks. Certified organic standards in Australia exclude engineered nanoparticles. It appears likely that other organic certifiers will also follow suit. The Soil Association was also the first to declare organic standards free from genetic engineering.

## Technical Aspects

### Size

Regulation of nanotechnology will require a definition of the size, in which particles and processes are recognized as operating at the nano-scale. The size-defining characteristic of nanotechnology is the subject of significant debate, and varies to include particles and materials in the scale of at least 100 to 300 nanometers (nm). Friends of the Earth Australia recommend defining nano-particles up to 300 nanometers (nm) in size. They argue that "particles up to a few hundred nanometers in size share many of the novel biological behaviors of nano-particles, including novel toxicity risks", and that "nano-materials up to approximately 300 nm in size can be taken up by individual cells". The UK Soil Association define nanotechnology to include manufactured nano-particles where the mean particle size is 200 nm or smaller. The U.S. National Nanotechnology Initiative defines nanotechnology as "the understanding and control of matter at dimensions of roughly 1 to 100 nm.

### Mass Thresholds

Regulatory frameworks for chemicals tend to be triggered by mass thresholds. This is certainly the case for the management of toxic chemicals in Australia through the National pollutant inventory. However, in the case of nanotechnology, nano-particle applications are unlikely to exceed these thresholds (tonnes/kilograms) due to the size and weight of nano-particles. As such, the Woodrow Wilson International Center for Scholars questions the usefulness of regulating nanotechnologies on the basis of their size/weight alone. They argue, for example, that the toxicity of nano-particles is more related to surface area than weight, and that emerging regulations should also take account of such factors.

# Molecular Nanotechnology

Molecular nanotechnology (MNT) is a technology based on the ability to build structures to complex, atomic specifications by means of mechanosynthesis. This is distinct from nanoscale materials. Based on Richard Feynman's vision of miniature factories using nanomachines to build complex products (including additional nanomachines), this advanced form of nanotechnology (or *molecular manufacturing*) would make use of positionally-controlled mechanosynthesis guided by molecular machine systems. MNT would involve combining physical principles demonstrated by biophysics, chemistry, other nanotechnologies, and the molecular machinery of life with the systems engineering principles found in modern macroscale factories.

## Introduction

While conventional chemistry uses inexact processes obtaining inexact results, and biology exploits inexact processes to obtain definitive results, molecular nanotechnology would employ original definitive processes to obtain definitive results. The desire in molecular nanotechnology would be to balance molecular reactions in positionally-controlled locations and orientations to obtain desired chemical reactions, and then to build systems by further assembling the products of these reactions.

A roadmap for the development of MNT is an objective of a broadly based technology project led by Battelle (the manager of several U.S. National Laboratories) and the Foresight Institute. The roadmap was originally scheduled for completion by late 2006, but was released in January 2008. The Nanofactory Collaboration is a more focused ongoing effort involving 23 researchers from 10 organizations and 4 countries that is developing a practical research agenda specifically aimed at positionally-controlled diamond mechanosynthesis and diamondoid nanofactory development. In August 2005, a task force consisting of 50+ international experts from various fields was organized by the Center for Responsible Nanotechnology to study the societal implications of molecular nanotechnology.

## Projected Applications and Capabilities

### Smart Materials and Nanosensors

One proposed application of MNT is so-called smart materials. This term refers to any sort of material designed and engineered at the nanometer scale for a specific task. It encompasses a wide variety of possible commercial applications. One example would be materials designed to respond differently to various molecules; such a capability could lead, for example, to artificial drugs which would recognize and render inert specific viruses. Another is the idea of self-healing structures, which would repair small tears in a surface naturally in the same way as self-sealing tires or human skin.

A MNT nanosensor would resemble a smart material, involving a small component within a larger machine that would react to its environment and change in some fundamental, intentional way. A very simple example: a photosensor might passively measure the incident light and discharge its absorbed energy as electricity when the light passes above or below a specified threshold, sending a signal to a larger machine. Such a sensor would supposedly cost less and use less power than a conventional sensor, and yet function usefully in all the same applications — for example, turning on parking lot lights when it gets dark.

While smart materials and nanosensors both exemplify useful applications of MNT, they pale in comparison with the complexity of the technology most popularly associated with the term: the replicating nanorobot.

## Replicating Nanorobots

MNT nanofacturing is popularly linked with the idea of swarms of coordinated nanoscale robots working together, a popularization of an early proposal by K. Eric Drexler in his 1986 discussions of MNT, but superseded in 1992. In this early proposal, sufficiently capable nanorobots would construct more nanorobots in an artificial environment containing special molecular building blocks.

Critics have doubted both the feasibility of self-replicating nanorobots and the feasibility of control if self-replicating nanorobots could be achieved: they cite the possibility of mutations removing any control and favoring reproduction of mutant pathogenic variations. Advocates address the first doubt by pointing out that the first macroscale autonomous machine replicator, made of Lego blocks, was built and operated experimentally in 2002. While there are sensory advantages present at the macroscale compared to the limited sensorium available at the nanoscale, proposals for positionally controlled nanoscale mechanosynthetic fabrication systems employ dead reckoning of tooltips combined with reliable reaction sequence design to ensure reliable results, hence a limited sensorium is no handicap; similar considerations apply to the positional assembly of small nanoparts. Advocates address the second doubt by arguing that bacteria are (of necessity) evolved to evolve, while nanorobot mutation could be actively prevented by common error-correcting techniques. Similar ideas are advocated in the Foresight Guidelines on Molecular Nanotechnology, and a map of the 137-dimensional replicator design space recently published by Freitas and Merkle provides numerous proposed methods by which replicators could, in principle, be safely controlled by good design.

However, the concept of suppressing mutation raises the question: How can design evolution occur at the nanoscale without a process of random mutation and deterministic selection? Critics argue that MNT advocates have not provided a substitute for such a process of evolution in this nanoscale arena where conventional sensory-based selection processes are lacking. The limits of the sensorium available at the nanoscale could make it difficult or impossible to winnow successes from failures. Advocates argue that design evolution should occur deterministically and strictly under human control, using the conventional engineering paradigm of modeling, design, prototyping, testing, analysis, and redesign.

In any event, since 1992 technical proposals for MNT do not include self-replicating nanorobots, and recent ethical guidelines put forth by MNT advocates prohibit unconstrained self-replication.

## Medical Nanorobots

One of the most important applications of MNT would be medical nanorobotics or nanomedicine, an area pioneered by Robert Freitas in numerous books and papers. The ability to design, build, and deploy large numbers of medical nanorobots would, at a minimum, make possible the rapid elimination of disease and the reliable and relatively painless recovery from physical trauma. Medical nanorobots might also make possible the convenient correction of genetic defects, and help to ensure a greatly expanded lifespan. More controversially, medical nanorobots might be used to augment natural human capabilities.

## Utility Fog

Diagram of a 100 micrometer foglet

Another proposed application of molecular nanotechnology is "utility fog" — in which a cloud of networked microscopic robots (simpler than assemblers) would change its shape and properties to form macroscopic objects and tools in accordance with software commands. Rather than modify the current practices of consuming material goods in different forms, utility fog would simply replace many physical objects.

## Phased-array Optics

Yet another proposed application of MNT would be phased-array optics (PAO). However, this appears to be a problem addressable by ordinary nanoscale technology.

PAO would use the principle of phased-array millimeter technology but at optical wavelengths. This would permit the duplication of any sort of optical effect but virtually. Users could request holograms, sunrises and sunsets, or floating lasers as the mood strikes. PAO systems were described in BC Crandall's *Nanotechnology: Molecular Speculations on Global Abundance* in the Brian Wowk article "Phased-Array Optics."

## Potential Social Impacts

Molecular manufacturing is a potential future subfield of nanotechnology that would make it possible to build complex structures at atomic precision. Molecular manufacturing requires significant advances in nanotechnology, but once achieved could produce highly advanced products at low costs and in large quantities in nanofactories weighing a kilogram or more. When nanofactories gain the ability to produce other nanofactories production may only be limited by relatively abundant factors such as input materials, energy and software.

The products of molecular manufacturing could range from cheaper, mass-produced versions of known high-tech products to novel products with added capabilities in many areas of application. Some applications that have been suggested are advanced smart materials, nanosensors, medical nanorobots and space travel. Additionally, molecular manufacturing could be used to cheaply produce highly advanced, durable weapons, which is an area of special concern regarding the impact of nanotechnology. Being equipped with compact computers and motors these could be increasingly autonomous and have a large range of capabilities.

According to Chris Phoenix and Mike Treder from the Center for Responsible Nanotechnology as well as Anders Sandberg from the Future of Humanity Institute molecular manufacturing is the application of nanotechnology that poses the most significant global catastrophic risk. Several nanotechnology researchers state that the bulk of risk from nanotechnology comes from the potential to lead to war, arms races and destructive global government. Several reasons have been suggested why the availability of nanotech weaponry may with significant likelihood lead to unstable arms races (compared to e.g. nuclear arms races): (1) A large number of players may be tempted to enter the race since the threshold for doing so is low; (2) the ability to make weapons with molecular manufacturing will be cheap and easy to hide; (3) therefore lack of insight into the other parties' capabilities can tempt players to arm out of caution or to launch preemptive strikes; (4) molecular manufacturing may reduce dependency on international trade, a potential peace-promoting factor; (5) wars of aggression may pose a smaller economic threat to the aggressor since manufacturing is cheap and humans may not be needed on the battlefield.

Since self-regulation by all state and non-state actors seems hard to achieve, measures to mitigate war-related risks have mainly been proposed in the area of international cooperation. International infrastructure may be expanded giving more sovereignty to the international level. This could help coordinate efforts for arms control. International institutions dedicated specifically to nanotechnology (perhaps analogously to the International Atomic Energy Agency IAEA) or general arms control may also be designed. One may also jointly make differential technological progress on defensive technologies, a policy that players should usually favour. The Center for Responsible Nanotechnology also suggest some technical restrictions. Improved transparency regarding technological capabilities may be another important facilitator for arms-control.

A grey goo is another catastrophic scenario, which was proposed by Eric Drexler in his 1986 book *Engines of Creation*, has been analyzed by Freitas in "Some Limits to Global Ecophagy by Biovorous Nanoreplicators, with Public Policy Recommendations" and has been a theme in mainstream media and fiction. This scenario involves tiny self-replicating robots that consume the entire biosphere using it as a source of energy and building blocks. Nanotech experts including Drexler now discredit the scenario. According to Chris Phoenix a "So-called grey goo could only be the product of a deliberate and difficult engineering process, not an accident". With the advent of nano-bio-tech, a different scenario called green goo has been forwarded. Here, the malignant substance is not nanobots but rather self-replicating biological organisms engineered through nanotechnology.

## Benefits

Nanotechnology (or molecular nanotechnology to refer more specifically to the goals discussed here) will let us continue the historical trends in manufacturing right up to the fundamental limits imposed by physical law. It will let us make remarkably powerful molecular computers. It will let us make materials over fifty times lighter than steel or aluminium alloy but with the same strength. We'll be able to make jets, rockets, cars or even chairs that, by today's standards, would be remarkably light, strong, and inexpensive. Molecular surgical tools, guided by molecular computers and injected into the blood stream could find and destroy cancer cells or invading bacteria, unclog arteries, or provide oxygen when the circulation is impaired.

Nanotechnology will replace our entire manufacturing base with a new, radically more precise, radically less expensive, and radically more flexible way of making products. The aim is not simply to replace today's computer chip making plants, but also to replace the assembly lines for cars, televisions, telephones, books, surgical tools, missiles, bookcases, airplanes, tractors, and all the rest. The objective is a pervasive change in manufacturing, a change that will leave virtually no product untouched. Economic progress and military readiness in the 21st Century will depend fundamentally on maintaining a competitive position in nanotechnology.

Despite the current early developmental status of nanotechnology and molecular nanotechnology, much concern surrounds MNT's anticipated impact on economics and on law. Whatever the exact effects, MNT, if achieved, would tend to reduce the scarcity of manufactured goods and make many more goods (such as food and health aids) manufacturable.

It is generally considered that future citizens of a molecular-nanotechnological society would still need money, in the form of unforgeable digital cash or physical specie (in special circumstances). They might use such money to buy goods and services that are unique, or limited within the solar system. These might include: matter, energy, information, real estate, design services, entertainment services, legal services, fame, political power, or the attention of other people to one's political/religious/philosophical message. Furthermore, futurists must consider war, even between prosperous states, and non-economic goals.

If MNT were realized, some resources would remain limited, because unique physical objects are limited (a plot of land in the real Jerusalem, mining rights to the larger near-earth asteroids) or because they depend on the goodwill of a particular person (the love of a famous person, a live audience in a musical concert). Demand will always exceed supply for some things, and a political economy may continue to exist in any case. Whether the interest in these limited resources would diminish with the advent of virtual reality, where they could be easily substituted, is yet unclear. One reason why it might not is a hypothetical preference for "the real thing", although such an opinion could easily be mollified if virtual reality were to develop to a certain level of quality.

MNT should make possible nanomedical capabilities able to cure any medical condition not already cured by advances in other areas. Good health would be common, and poor health of any form would be as rare as smallpox and scurvy are today. Even cryonics would be feasible, as cryopreserved tissue could be fully repaired.

## Risks

Molecular nanotechnology is one of the technologies that some analysts believe could lead to a Technological Singularity. Some feel that molecular nanotechnology would have daunting risks. It conceivably could enable cheaper and more destructive conventional weapons. Also, molecular nanotechnology might permit weapons of mass destruction that could self-replicate, as viruses and cancer cells do when attacking the human body. Commentators generally agree that, in the event molecular nanotechnology were developed, its self-replication should be permitted only under very controlled or "inherently safe" conditions.

A fear exists that nanomechanical robots, if achieved, and if designed to self-replicate using naturally occurring materials (a difficult task), could consume the entire planet in their hunger for raw materials, or simply crowd out natural life, out-competing it for energy (as happened historically when blue-green algae appeared and outcompeted earlier life forms). Some commentators have referred to this situation as the "grey goo" or "ecophagy" scenario. K. Eric Drexler considers an accidental "grey goo" scenario extremely unlikely and says so in later editions of *Engines of Creation*.

In light of this perception of potential danger, the Foresight Institute (founded by K. Eric Drexler to prepare for the arrival of future technologies) has drafted a set of guidelines for the ethical development of nanotechnology. These include the banning of free-foraging self-replicating pseudo-organisms on the Earth's surface, at least, and possibly in other places.

## Technical Issues and Criticism

The feasibility of the basic technologies analyzed in *Nanosystems* has been the subject of a formal scientific review by U.S. National Academy of Sciences, and has also been the focus of extensive debate on the internet and in the popular press.

## Study and Recommendations by the U.S. National Academy of Sciences

In 2006, U.S. National Academy of Sciences released the report of a study of molecular manufacturing as part of a longer report, *A Matter of Size: Triennial Review of the National Nanotechnology Initiative* The study committee reviewed the technical content of *Nanosystems*, and in its conclusion states that no current theoretical analysis can be considered definitive regarding several questions of potential system performance, and that optimal paths for implementing high-performance systems cannot be predicted with confidence. It recommends experimental research to advance knowledge in this area:

> "Although theoretical calculations can be made today, the eventually attainable range of chemical reaction cycles, error rates, speed of operation, and thermodynamic efficiencies of such bottom-up manufacturing systems cannot be reliably predicted at this time. Thus, the eventually attainable perfection and complexity of manufactured products, while they can be calculated in theory, cannot be predicted with confidence. Finally, the optimum research paths that might lead to systems which greatly exceed the thermodynamic efficiencies and other capabilities of biological systems cannot be reliably predicted at this time. Research funding that is based on the ability of investigators to produce experimental demonstrations that link to abstract models and guide long-term vision is most appropriate to achieve this goal."

## Assemblers Versus Nanofactories

A section heading in Drexler's *Engines of Creation* reads "Universal Assemblers", and the following text speaks of multiple types of assemblers which, collectively, could hypothetically "build almost anything that the laws of nature allow to exist." Drexler's colleague Ralph Merkle has noted that, contrary to widespread legend, Drexler never claimed that assembler systems could build absolutely any molecular structure. The endnotes in Drexler's book explain the qualification "almost": "For example, a delicate structure might be designed that, like a stone arch, would self-destruct unless all its pieces were already in place. If there were no room in the design for the placement and removal of a scaffolding, then the structure might be impossible to build. Few structures of practical interest seem likely to exhibit such a problem, however."

In 1992, Drexler published *Nanosystems: Molecular Machinery, Manufacturing, and Computation*, a detailed proposal for synthesizing stiff covalent structures using a table-top factory. Diamondoid structures and other stiff covalent structures, if achieved, would have a wide range of possible applications, going far beyond current MEMS technology. An outline of a path was put forward in 1992 for building a table-top factory in the absence of an assembler. Other researchers have begun advancing tentative, alternative proposed paths for this in the years since Nanosystems was published.

## Hard Versus Soft Nanotechnology

In 2004 Richard Jones wrote Soft Machines (nanotechnology and life), a book for lay audiences published by Oxford University. In this book he describes radical nanotechnology (as advocated by Drexler) as a deterministic/mechanistic idea of nano engineered machines that does not take into account the nanoscale challenges such as wetness, stickness, Brownian motion, and high viscosity. He also explains what is soft nanotechnology or more appropriatelly biomimetic nanotechnology which is the way forward, if not the best way, to design functional nanodevices that can cope with all the problems at a nanoscale. One can think of soft nanotechnology as the development of nanomachines that uses the lessons learned from biology on how things work, chemistry to precisely engineer such devices and stochastic physics to model the system and its natural processes in detail.

## The Smalley-drexler Debate

Several researchers, including Nobel Prize winner Dr. Richard Smalley (1943–2005), attacked the notion of universal assemblers, leading to a rebuttal from Drexler and colleagues, and eventually to an exchange of letters. Smalley argued that chemistry is extremely complicated, reactions are hard to control, and that a universal assembler is science fiction. Drexler and colleagues, however, noted that Drexler never proposed universal assemblers able to make absolutely anything, but instead proposed more limited assemblers able to make a very wide variety of things. They challenged the relevance of Smalley's arguments to the more specific proposals advanced in *Nanosystems*. Also, Smalley argued that nearly all of modern chemistry involves reactions that take place in a solvent (usually water), because the small molecules of a solvent contribute many things, such as lowering binding energies for transition states. Since nearly all known chemistry requires a solvent, Smalley felt that Drexler's proposal to use a high vacuum environment was not feasible. However, Drexler addresses this in Nanosystems by showing mathematically that well designed catalysts can provide the effects of a solvent and can fundamentally be made even more efficient than a solvent/enzyme reaction could ever be. It is noteworthy that, contrary to Smalley's opinion that enzymes require water, "Not only do enzymes work vigorously in anhydrous organic media, but in this unnatural milieu they acquire remarkable properties such as greatly enhanced stability, radically altered substrate and enantiomeric specificities, molecular memory, and the ability to catalyse unusual reactions.""*Enzymatic catalysis in anhydrous organic solvents.*". April 1989. ""*Enzymatic catalysis in anhydrous organic solvents*" (PDF). April 1989.*

## Design Issues

For the future, some means have to be found for MNT design evolution at the nanoscale which mimics the process of biological evolution at the molecular scale. Biological evolution proceeds by random variation in ensemble averages of organisms combined with culling of the less-successful variants and reproduction of the more-successful variants, and macroscale engineering design also proceeds by a process of design evolution from simplicity to complexity as set forth somewhat satirically by John Gall: "A complex system that works is invariably found to have evolved from a simple system that worked. . . . A complex system designed from scratch never works and can not be patched up to make it work. You have to start over, beginning with a system that works." A breakthrough in MNT is needed which proceeds from the simple atomic ensembles which can be built with, e.g., an STM to complex MNT systems via a process of design evolution. A handicap in this process is the difficulty of seeing and manipulation at the nanoscale compared to the macroscale which makes deterministic selection of successful trials difficult; in contrast biological evolution proceeds via action of what Richard Dawkins has called the "blind watchmaker" comprising random molecular variation and deterministic reproduction/extinction.

At present in 2007 the practice of nanotechnology embraces both stochastic approaches (in which, for example, supramolecular chemistry creates waterproof pants) and deterministic approaches wherein single molecules (created by stochastic chemistry) are manipulated on substrate surfaces (created by stochastic deposition methods) by deterministic methods comprising nudging them with STM or AFM probes and causing simple binding or cleavage reactions to occur. The dream of a complex, deterministic molecular nanotechnology remains elusive. Since the mid-1990s, thousands of surface scientists and thin film technocrats have latched on to the nanotechnology bandwagon and redefined their disciplines as nanotechnology. This has caused much confusion in the field and has spawned thousands of "nano"-papers on the peer reviewed literature. Most of these reports are extensions of the more ordinary research done in the parent fields.

# The Feasibility of the Proposals in Nanosystems

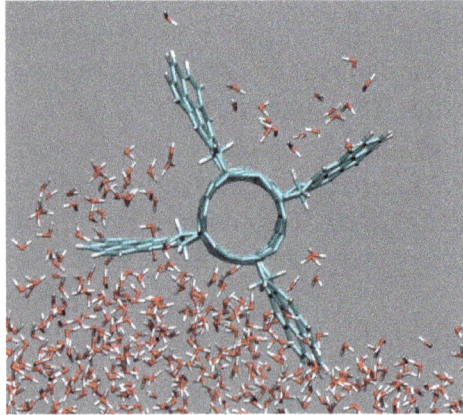

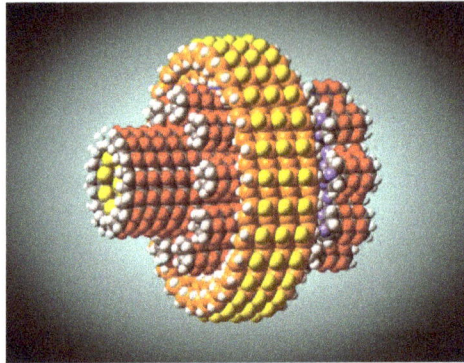

Top, a molecular propellor. Bottom, a molecular planetary gear system. The feasibility of devices like these has been questioned.

The feasibility of Drexler's proposals largely depends, therefore, on whether designs like those in *Nanosystems* could be built in the absence of a universal assembler to build them and would work as described. Supporters of molecular nanotechnology frequently claim that no significant errors have been discovered in *Nanosystems* since 1992. Even some critics concede that "Drexler has carefully considered a number of physical principles underlying the 'high level' aspects of the nanosystems he proposes and, indeed, has thought in some detail" about some issues.

Other critics claim, however, that *Nanosystems* omits important chemical details about the low-level 'machine language' of molecular nanotechnology. They also claim that much of the other low-level chemistry in *Nanosystems* requires extensive further work, and that Drexler's higher-level designs therefore rest on speculative foundations. Recent such further work by Freitas and Merkle is aimed at strengthening these foundations by filling the existing gaps in the low-level chemistry.

Drexler argues that we may need to wait until our conventional nanotechnology improves before solving these issues: "Molecular manufacturing will result from a series of advances in molecular machine systems, much as the first Moon landing resulted from a series of advances in liquid-fuel rocket systems. We are now in a position like that of the British Interplanetary Society of the 1930s which described how multistage liquid-fueled rockets could reach the Moon and pointed to early rockets as illustrations of the basic principle." However, Freitas and Merkle argue that a focused effort to achieve diamond mechanosynthesis (DMS) can begin now, using existing technology, and might achieve success in less than a decade if their "direct-to-DMS approach is pursued rather than a more circuitous development approach that seeks to implement less efficacious nondiamondoid molecular manufacturing technologies before progressing to diamondoid".

To summarize the arguments against feasibility: First, critics argue that a primary barrier to achieving molecular nanotechnology is the lack of an efficient way to create machines on a molecular/atomic scale, especially in the absence of a well-defined path toward a self-replicating assembler or diamondoid nanofactory. Advocates respond that a preliminary research path leading to a diamondoid nanofactory is being developed.

A second difficulty in reaching molecular nanotechnology is design. Hand design of a gear or bearing at the level of atoms might take a few to several weeks. While Drexler, Merkle and others have created designs of simple parts, no comprehensive design effort for anything approaching the complexity of a Model T Ford has been attempted. Advocates respond that it is difficult to undertake a comprehensive design effort in the absence of significant funding for such efforts, and that despite this handicap much useful design-ahead has nevertheless been accomplished with new software tools that have been developed, e.g., at Nanorex.

In the latest report *A Matter of Size: Triennial Review of the National Nanotechnology Initiative* put out by the National Academies Press in December 2006 (roughly twenty years after Engines of Creation was published), no clear way forward toward molecular nanotechnology could yet be seen, as per the conclusion on page 108 of that report: "Although theoretical calculations can be made today, the eventually attainable range of chemical reaction cycles, error rates, speed of operation, and thermodynamic efficiencies of such bottom-up manufacturing systems cannot be reliably predicted at this time. Thus, the eventually attainable perfection and complexity of manufactured products, while they can be calculated in theory, cannot be predicted with confidence. Finally, the optimum research paths that might lead to systems which greatly exceed the thermodynamic efficiencies and other capabilities of biological systems cannot be reliably predicted at this time. Research funding that is based on the ability of investigators to produce experimental demonstrations that link to abstract models and guide long-term vision is most appropriate to achieve this goal." This call for research leading to demonstrations is welcomed by groups such as the Nanofactory Collaboration who are specifically seeking experimental successes in diamond mechanosynthesis. The "Technology Roadmap for Productive Nanosystems" aims to offer additional constructive insights.

It is perhaps interesting to ask whether or not most structures consistent with physical law can in fact be manufactured. Advocates assert that to achieve most of the vision of molecular manufacturing it is not necessary to be able to build "any structure that is compatible with natural law." Rather, it is necessary to be able to build only a sufficient (possibly modest) subset of such structures—as is true, in fact, of any practical manufacturing process used in the world today, and is true even in biology. In any event, as Richard Feynman once said, "It is scientific only to say what's more likely or less likely, and not to be proving all the time what's possible or impossible."

## Existing Work on Diamond Mechanosynthesis

There is a growing body of peer-reviewed theoretical work on synthesizing diamond by mechanically removing/adding hydrogen atoms and depositing carbon atoms (a process known as mechanosynthesis). This work is slowly permeating the broader nanoscience community and is being critiqued. For instance, Peng et al. (2006) (in the continuing research effort by Freitas, Merkle and their collaborators) reports that the most-studied mechanosynthesis tooltip motif (DCB6Ge) successfully places a $C_2$ carbon dimer on a C(110) diamond surface at both 300 K (room temperature) and 80 K (liquid nitrogen temperature), and that the silicon variant (DCB6Si) also works at 80 K but not at 300 K. Over 100,000 CPU hours were invested in this latest study. The DCB6 tooltip motif, initially described by Merkle and Freitas at a Foresight Conference in 2002, was the first complete tooltip ever proposed for diamond mechanosynthesis and remains the only tooltip motif that has been successfully simulated for its intended function on a full 200-atom diamond surface.

The tooltips modeled in this work are intended to be used only in carefully controlled environments (e. g., vacuum). Maximum acceptable limits for tooltip translational and rotational misplacement errors are reported in Peng et al. (2006) -- tooltips must be positioned with great accuracy to avoid bonding the dimer incorrectly. Peng et al. (2006) reports that increasing the handle thickness from 4 support planes of C atoms above the tooltip to 5 planes decreases the resonance frequency of the entire structure from 2.0 THz to 1.8 THz. More importantly, the vibrational footprints of a DCB6Ge tooltip mounted on a 384-atom handle and of the same tooltip mounted on a similarly constrained but much larger 636-atom "crossbar" handle are virtually identical in the non-crossbar directions. Additional computational studies modeling still bigger handle structures are welcome, but the ability to precisely position SPM tips to the requisite atomic accuracy has been repeatedly demonstrated experimentally at low temperature, or even at room temperature constituting a basic existence proof for this capability.

Further research to consider additional tooltips will require time-consuming computational chemistry and difficult laboratory work.

A working nanofactory would require a variety of well-designed tips for different reactions, and detailed analyses of placing atoms on more complicated surfaces. Although this appears a challenging problem given current resources, many tools will be available to help future researchers: Moore's Law predicts further increases in computer power, semiconductor fabrication techniques continue to approach the nanoscale, and researchers grow ever more skilled at using proteins, ribosomes and DNA to perform novel chemistry.

## Works of Fiction

- In The Diamond Age by Neal Stephenson diamond can be constructed by simply building it out of carbon atoms. Also all sorts of devices from dust size detection devices to giant diamond zeppelins are constructed atom by atom using only carbon, oxygen, nitrogen and chlorine atoms.

- In the novel *Tomorrow* by Andrew Saltzman (ISBN 1-4243-1027-X), a scientist uses nanorobotics to create a liquid that when inserted into the bloodstream, renders one nearly invincible given that the microscopic machines repair tissue almost instantaneously after it is damaged.

- In the roleplaying game Splicers by Palladium Books, humanity has succumbed to a "nanobot plague" that causes any object made of a non-precious metal to twist and change shape (sometimes into a type of robot) moments after being touched by a human. The object will then proceed to attack the human. This has forced humanity to develop "biotechnological" devices to replace those previously made of metal.

- On the television show Mystery Science Theater 3000, the Nanites (voiced variously by Kevin Murphy, Paul Chaplin, Mary Jo Pehl, and Bridget Jones) - are self-replicating, bio-engineered organisms that work on the ship, they are microscopic creatures that reside in the Satellite of Love's computer systems. (They are similar to the creatures in *Star Trek: The Next Generation* episode "Evolution", which featured "nanites" taking over the *Enterprise*.) The Nanites made their first appearance in season 8. Based on the concept of nanotechnology, their comical *deus ex machina* activities included such diverse tasks as instant repair and construction, hairstyling, performing a Nanite variation of a flea circus, conducting a microscopic war, and even destroying the Observers' planet after a dangerously vague request from Mike to "take care of [a] little problem". They also ran a microbrewery.

## Societal Impact of Nanotechnology

The societal impact of nanotechnology are the potential benefits and challenges that the introduction of novel nanotechnological devices and materials may hold for society and human interaction. The term is sometimes expanded to also include nanotechnology's health and environmental impact, but this article will only consider the social and political impact of nanotechnology.

As nanotechnology is an emerging field and most of its applications are still speculative, there is much debate about what positive and negative effects that nanotechnology might have.

## Overview

Beyond the toxicity risks to human health and the environment which are associated with first-generation nanomaterials, nanotechnology has broader societal implications and poses broader social challenges. Social scientists have suggested that nanotechnology's social issues should be understood and assessed not simply as "downstream" risks or impacts. Rather, the challenges should be factored into "upstream" research and decision making in order to ensure technology development that meets social objectives.

Many social scientists and organizations in civil society suggest that technology assessment and governance should also involve public participation.

Some observers suggest that nanotechnology will build incrementally, as did the 18-19th century industrial revolution, until it gathers pace to drive a nanotechnological revolution that will radically reshape our economies, our labor markets, international trade, international relations, social structures, civil liberties, our relationship with the natural world and even what we understand to be human. Others suggest that it may be more accurate to describe change driven by nanotechnology as a "technological tsunami". Just like a tsunami, analysts warn that rapid nanotechnology-driven change will necessarily have profound disruptive impacts. As the APEC Center for Technology Foresight observes:

If nanotechnology is going to revolutionize manufacturing, health care, energy supply, communications and probably defense, then it will transform labour and the workplace, the medical system, the transportation and power infrastructures and the military. None of these latter will be changed without significant social disruption.

Those concerned with the negative impact of nanotechnology suggest that it will simply exacerbate problems stemming from existing socio-economic inequity and unequal distributions of power, creating greater inequities between rich and poor through an inevitable nano-divide (the gap between those who control the new nanotechnologies and those whose products, services or labour are displaced by them). Analysts suggest the possibility that nanotechnology has the potential to destabilize international relations through a nano arms race and the increased potential for bioweaponry; thus, providing the tools for ubiquitous surveillance with significant implications for civil liberties. Also, many critics believe it might break down the barriers between life and non-life through nanobiotechnology, redefining even what it means to be human.

Nanoethicists posit that such a transformative technology could exacerbate the divisions of rich and poor – the so-called "nano divide." However nanotechnology makes the production of technology, e.g. computers, cellular phones, health technology etcetera, cheaper and therefore accessible to the poor.

In fact, many of the most enthusiastic proponents of nanotechnology, such as transhumanists, see the nascent science as a mechanism to changing human nature itself – going beyond curing disease and enhancing human characteristics. Discussions on nanoethics have been hosted by the federal government, especially in the context of "converging technologies" – a catch-phrase used to refer to nano, biotech, information technology, and cognitive science.

## Possible Military Applications

Possible military applications of nanotechnology have been suggested in the fields of soldier enhancement and chemical weapons amongst others. However, more socially disruptive weapon systems are to be expected from molecular manufacturing, a potential future form of nanotechnology that would make it possible to build complex structures at atomic precision. Molecular manufacturing requires significant advances in nanotechnology, but its supporters posit that once achieved it could produce highly advanced products at low costs and in large quantities in nanofactories weighing a kilogram or more. If nanofactories gain the ability to produce other nanofactories production may only be limited by relatively abundant factors such as input materials, energy and software.

Molecular manufacturing might be used to cheaply produce, among many other products, highly advanced, durable weapons. Being equipped with compact computers and motors these might be increasingly autonomous and have a large range of capabilities.

According to Chris Phoenix and Mike Treder from the Center for Responsible Nanotechnology as well as Anders Sandberg from the Future of Humanity Institute the military uses of molecular manufacturing are the applications of nanotechnology that pose the most significant global catastrophic risk. Several nanotechnology researchers state that the bulk of risk from nanotechnology comes from the potential to lead to war, arms races and destructive global government. Several reasons have been suggested why the availability of nanotech weaponry may with significant likelihood lead to unstable arms races (compared to e.g. nuclear arms races): (1) A large number of players may be tempted to enter the race since the threshold for doing so is low; (2) the ability to make weapons with molecular manufacturing might be cheap and easy to hide; (3) therefore lack of insight into the other parties' capabilities can tempt players to arm out of caution or to launch preemptive strikes; (4) molecular manufacturing may reduce dependency on international trade, a potential peace-promoting factor; (5) wars of aggression may pose a smaller economic threat to the aggressor since manufacturing is cheap and humans may not be needed on the battlefield.

Self-regulation by all state and non-state actors has been called hard to achieve, so measures to mitigate war-related risks have mainly been proposed in the area of international cooperation. International infrastructure may be expanded giving more sovereignty to the international level. This could help coordinate efforts for arms control. Some have put forth that international institutions dedicated specifically to nanotechnology (perhaps analogously to the International Atomic Energy Agency IAEA) or general arms control may also be designed. One may also jointly make differential technological progress on defensive technologies. The Center for Responsible Nanotechnology also suggest some technical restrictions. Improved transparency regarding technological capabilities may be another important facilitator for arms-control.

## Intellectual Property Issues

On the structural level, critics of nanotechnology point to a new world of ownership and corporate control opened up by nanotechnology. The claim is that, just as biotechnology's ability to manipulate genes went hand in hand with the patenting of life, so too nanotechnology's ability to manipulate molecules has led to the patenting of matter. The last few years has seen a gold rush to claim patents at the nanoscale. Academics have warned that the resultant patent thicket is harming progress in the technology and have argued in the top journal *Nature* that there should be a moratorium on patents on "building block" nanotechnologies. Over 800 nano-related patents were granted in 2003, and the numbers are increasing year to year. Corporations are already taking out broad-ranging patents on nanoscale discoveries and inventions. For example, two corporations, NEC and IBM, hold the basic patents on carbon nanotubes, one of the current cornerstones of nanotechnology. Carbon nanotubes have a wide range of uses, and look set to become crucial to several industries from electronics and computers, to strengthened materials to drug delivery and diagnostics. Carbon nanotubes are poised to become a major traded commodity with the potential to replace major conventional raw materials. However, as their use expands, anyone seeking to (legally) manufacture or sell carbon nanotubes, no matter what the application, must first buy a license from NEC or IBM.

The United States' essential facilities doctrine may be of importance as well as other anti-trust laws.

## Potential Benefits and Risks for Developing Countries

Nanotechnologies may provide new solutions for the millions of people in developing countries who lack access to basic services, such as safe water, reliable energy, health care, and education. The United Nations has set Millennium Development Goals for meeting these needs. The 2004 UN Task Force on Science, Technology and Innovation noted that some of the advantages of nanotechnology include production using little labor, land, or maintenance, high productivity, low cost, and modest requirements for materials and energy.

Many developing countries, for example Costa Rica, Chile, Bangladesh, Thailand, and Malaysia, are investing considerable resources in research and development of nano-technologies. Emerging economies such as Brazil, China, India and South Africa are spending millions of US dollars annually on R&D, and are rapidly increasing their scientific output as demonstrated by their increasing numbers of publications in peer-reviewed scientific publications.

Potential opportunities of nanotechnologies to help address critical international development priorities include improved water purification systems, energy systems, medicine and pharmaceuticals, food production and nutrition, and information and communications technologies. Nanotechnologies are already incorporated in products that are on the market. Other nanotechnologies are still in the research phase, while others are concepts that are years or decades away from development.

Applying nanotechnologies in developing countries raises similar questions about the environmental, health, and societal risks described in the previous section. Additional challenges have been raised regarding the linkages between nanotechnology and development.

Protection of the environment, human health and worker safety in developing countries often suffers from a combination of factors that can include but are not limited to lack of robust environmental, human health, and worker safety regulations; poorly or unenforced regulation which is linked to a lack of physical (e.g., equipment) and human capacity (i.e., properly trained regulatory staff). Often, these nations require assistance, particularly financial assistance, to develop the scientific and institutional capacity to adequately assess and manage risks, including the necessary infrastructure such as laboratories and technology for detection.

Very little is known about the risks and broader impacts of nanotechnology. At a time of great uncertainty over the impacts of nanotechnology it will be challenging for governments, companies, civil society organizations, and the general public in developing countries, as in developed countries, to make decisions about the governance of nanotechnology.

Companies, and to a lesser extent governments and universities, are receiving patents on nanotechnology. The rapid increase in patenting of nanotechnology is illustrated by the fact that in the US, there were 500 nanotechnology patent applications in 1998 and 1,300 in 2000. Some patents are very broadly defined, which has raised concern among some groups that the rush to patent could slow innovation and drive up costs of products, thus reducing the potential for innovations that could benefit low income populations in developing countries.

There is a clear link between commodities and poverty. Many least developed countries are dependent on a few commodities for employment, government revenue, and export earnings. Many applications of nanotechnology are being developed that could impact global demand for specific commodities. For instance, certain nanoscale materials could enhance the strength and durability of rubber, which might eventually lead to a decrease in demand for natural rubber. Other nanotechnology applications may result in increases in demand for certain commodities. For example, demand for titanium may increase as a result of new uses for nanoscale titanium oxides, such as titanium dioxide nanotubes that can be used to produce and store hydrogen for use as fuel. Various organizations have called for international dialogue on mechanisms that will allow developing countries to anticipate and proactively adjust to these changes.

In 2003, Meridian Institute began the Global Dialogue on Nanotechnology and the Poor: Opportunities and Risks (GDNP) to raise awareness of the opportunities and risks of nanotechnology for developing countries, close the gaps within and between sectors of society to catalyze actions that address specific opportunities and risks of nanotechnology for developing countries, and identify ways that science and technology can play an appropriate role in the development process. The GDNP has released several publicly accessible papers on nanotechnology and development, including "Nanotechnology and the Poor: Opportunities and Risks - Closing the Gaps Within and Between Sectors of Society"; "Nanotechnology, Water, and Development"; and "Overview and Comparison of Conventional and Nano-Based Water Treatment Technologies".

## Social Justice and Civil Liberties

Concerns are frequently raised that the claimed benefits of nanotechnology will not be evenly distributed, and that any benefits (including technical and/or economic) associated with nanotechnology will only reach affluent nations. The majority of nanotechnology research and development - and patents for nanomaterials and products - is concentrated in developed countries (including the United States, Japan, Germany, Canada and France). In addition, most patents related to nanotechnology are concentrated amongst few multinational corporations, including IBM, Micron Technologies, Advanced Micro Devices and Intel. This has led to fears that it will be unlikely that developing countries will have access to the infrastructure, funding and human resources required to support nanotechnology research and development, and that this is likely to exacerbate such inequalities.

Producers in developing countries could also be disadvantaged by the replacement of natural products (including rubber, cotton, coffee and tea) by developments in nanotechnology. These natural products are important export crops for developing countries, and many farmers' livelihoods depend on them. It has been argued that their substitution with industrial nano-products could negatively impact the economies of developing countries, that have traditionally relied on these export crops.

It is proposed that nanotechnology can only be effective in alleviating poverty and aid development "when adapted to social, cultural and local institutional contexts, and chosen and designed with the active participation by citizens right from the commencement point" (Invernizzi et al. 2008, p. 132).

## Effects on Laborers

Ray Kurzweil has speculated in *The Singularity is Near* that people who work in unskilled labor jobs for a livelihood may become the first human workers to be displaced by the constant use of nanotechnology in the workplace, noting that layoffs often affect the jobs based around the lowest technology level before attacking jobs with the highest technology level possible. It has been noted that every major economic era has stimulated a global revolution both in the kinds of jobs that are available to people and the kind of training they need to achieve these jobs, and there is concern that the world's educational systems have lagged behind in preparing students for the "Nanotech Age".

It has also been speculated that nanotechnology may give rise to nanofactories which may have superior capabilities to conventional factories due to their small carbon and physical footprint on the global and regional environment. The miniaturization and transformation of the multi-acre conventional factory into the nanofactory may not interfere with their ability to deliver a high quality product; the product may be of even greater quality due to the lack of human errors in the production stages. Nanofactory systems may use precise atomic precisioning and contribute to making superior quality products that the "bulk chemistry" method used in 20th century and early 21st currently cannot produce. These advances might shift the computerized workforce in an even more complex direction, requiring skills in genetics, nanotechnology, and robotics.

## References

- Drexler, K. Eric (1992). Nanosystems: Molecular Machinery, Manufacturing, and Computation. New York: John Wiley & Sons. ISBN 0-471-57547-X.

- Allhoff, Fritz; Lin, Patrick; Moore, Daniel (2010). What is nanotechnology and why does it matter?: from science to ethics. John Wiley and Sons. pp. 3–5. ISBN 1-4051-7545-1.

- Petty, M.C.; Bryce, M.R. & Bloor, D. (1995). Introduction to Molecular Electronics. New York: Oxford University Press. pp. 1–25. ISBN 0-19-521156-1.

- Chris Phoenix; Mike Treder (2008). "Chapter 21: Nanotechnology as global catastrophic risk". In Bostrom, Nick; Cirkovic, Milan M. Global catastrophic risks. Oxford: Oxford University Press. ISBN 978-0-19-857050-9.

- R. V. Lapshin (2011). "Feature-oriented scanning probe microscopy". In H. S. Nalwa. Encyclopedia of Nanoscience and Nanotechnology (PDF). 14. USA: American Scientific Publishers. pp. 105–115. ISBN 1-58883-163-9.

- "About the National Nanotechnology Initiative". United States National Nanotechnology Initiative. 2016. Retrieved 4 June 2016.

- Smith, Erin Geiger (14 February 2013). "U.S.-based inventors lead world in nanotechnology patents: study". Technology. Reuters. Retrieved 4 June 2016.

- Prigg, Mark (2 October 2015). "The end of silicon? IBM reveals carbon nanotube breakthrough that could revolutionise computing and lead to ultrafast artificial intelligence 'brain chips'". Associated Newspapers Ltd, The Daily Mail. Retrieved 4 June 2016.

- "CDC – Nanotechnology – NIOSH Workplace Safety and Health Topic". National Institute for Occupational Safety and Health. June 15, 2012. Retrieved 2012-08-24.

- "CDC – NIOSH Publications and Products – Filling the Knowledge Gaps for Safe Nanotechnology in the Workplace". National Institute for Occupational Safety and Health. November 7, 2012. Retrieved 2012-11-08.

- "Current Intelligence Bulletin 63: Occupational Exposure to Titanium Dioxide" (PDF). United States National Institute for Occupational Safety and Health. Retrieved 2012-02-19.

- Zhuang Z, Viscusi D (December 7, 2011). "CDC - NIOSH Science Blog - Respiratory Protection for Workers Handling Engineering Nanoparticles". National Institute for Occupational Safety and Health. Retrieved 2012-08-24.

- "Nanotechnology Information Center: Properties, Applications, Research, and Safety Guidelines". American Elements. Retrieved 13 May 2011.

- "Analysis: This is the first publicly available on-line inventory of nanotechnology-based consumer products". The Project on Emerging Nanotechnologies. 2008. Retrieved 13 May 2011.

# Essential Elements of Nanoengineering

The essential elements of nanoengineering are nanofluids, nanomaterials, nanoparticles, nanometers, carbon nanotubes and nanosensors. Nanofluids are fluids that contain nanometer-sized particles whereas nanoparticles are materials of size between 1-100 nanoscale. The topics discussed in the chapter are of great importance to broaden the existing knowledge of nanoengineering.

## Nanofluid

A nanofluid is a fluid containing nanometer-sized particles, called nanoparticles. These fluids are engineered colloidal suspensions of nanoparticles in a base fluid. The nanoparticles used in nanofluids are typically made of metals, oxides, carbides, or carbon nanotubes. Common base fluids include water, ethylene glycol and oil.

Nanofluids have novel properties that make them potentially useful in many applications in heat transfer, including microelectronics, fuel cells, pharmaceutical processes, and hybrid-powered engines, engine cooling/vehicle thermal management, domestic refrigerator, chiller, heat exchanger,in grinding, machining and in boiler flue gas temperature reduction. They exhibit enhanced thermal conductivity and the convective heat transfer coefficient compared to the base fluid. Knowledge of the rheological behaviour of nanofluids is found to be very critical in deciding their suitability for convective heat transfer applications Nanofluids also have special acoustical properties and in ultrasonic fields display additional shear-wave reconversion of an incident compressional wave; the effect becomes more pronounced as concentration increases.

In analysis such as computational fluid dynamics (CFD), nanofluids can be assumed to be single phase fluids. However, almost all of new academic paper use two-phase assumption. Classical theory of single phase fluids can be applied, where physical properties of nanofluid is taken as a function of properties of both constituents and their concentrations. An alternative approach simulates nanofluids using a two-component model.

The spreading of a nanofluid droplet is enhanced by the solid-like ordering structure of nanoparticles assembled near the contact line by diffusion, which gives rise to a structural disjoining pressure in the vicinity of the contact line. However, such enhancement is not observed for small droplets with diameter of nanometer scale, because the wetting time scale is much smaller than the diffusion time scale.

## Synthesis

Nanofluids are produced by several techniques they are, 1.Direct Evaporation (1 step), 2.Gas condensation/dispersion (2 step), 3.Chemical vapour condensation (1 step), 4.Chemical precipitation (1 step). Several liquids including water, ethylene glycol, and oils have been used as base fluids. Although stabilization can be a challenge, on-going research indicates that it is possible. Nano-materials used so far in nanofluid synthesis include metallic particles, oxide particles, carbon nanotubes, graphene nano-flakes and ceramic particles.

## Smart Cooling Nanofluids

Realizing the modest thermal conductivity enhancement in conventional nanofluids, a team of researchers at Indira Gandhi Centre for Atomic Research Centre, Kalpakkam developed a new class of magnetically polarizable nanofluids where the thermal conductivity enhancement up to 300% of basefluids is demonstrated. Fatty-acid-capped magnetite nanoparticles of different sizes (3-10 nm) have been synthesized for this purpose. It has been shown that both the thermal and rheological properties of such magnetic nanofluids are tunable by varying the magnetic field strength and orientation with respect to the direction of heat flow. Such response stimuli fluids are reversibly switchable and have applications in miniature devices such as micro- and nano-electromechanical systems. In 2013, Azizian et al. considered the effect of an external magnetic field on the convective heat transfer coefficient of water-based magnetite nanofluid experimentally under laminar flow regime. Up to 300% enhancement obtained at Re=745 and magnetic field gradient of 32.5 mT/mm. The effect of the magnetic field on the pressure drop was not as significant.

## Nanoparticle Migration

In nanofluids, it is recognized that nanoparticles do not follow the fluid streamlines passively. In fact, there are some reasons that induce a slip velocity between the nanoparticles and the base fluid. Movements of nanoparticles has significant impact on rheological and thermophysical properties of the nanofluids. Therefore, investigating the nanoparticles motion is critical for evaluating the performance of nanoparticles inclusion to the base fluid as a heat transfer medium. Since the nanoparticles are very small ( 100 nm), Brownian and thermophoretic diffusivities are the main slip mechanisms in nanofluids, as Buongiorno declared. Brownian diffusion is due to random drifting of suspended nanoparticles in the base fluid which originates from continuous collisions among the nanoparticles and liquid molecules. Thermophoresis induces nanoparticle migration from warmer to colder region (in opposite direction of the temperature gradient), making a non-uniform nanoparticle volume fraction distribution.

In fact, theoretical models estimated that nanoparticles are non-homogeneously distributed. The level of non-uniformity is completely depend on thermal boundary conditions, the nanoparticle size, shape, and material. Rigorous readers encouraged to find more interesting results in open literature.

## Response Stimuli Nanofluids for Sensing Applications

Researchers have invented a nanofluid-based ultrasensitive optical sensor that changes its colour on exposure to extremely low concentrations of toxic cations. The sensor is useful in detecting minute traces of cations in industrial and environmental samples. Existing techniques for monitoring cations levels in industrial and environmental samples are expensive, complex and time-consuming. The sensor is designed with a magnetic nanofluid that consists of nano-droplets with magnetic grains suspended in water. At a fixed magnetic field, a light source illuminates the nanofluid where the colour of the nanofluid changes depending on the cation concentration. This color change occurs within a second after exposure to cations, much faster than other existing cation sensing methods.

Such response stimulus nanofluids are also used to detect and image defects in ferromagnetic components. The photonic eye, as it has been called, is based on a magnetically polarizable nano-emulsion that changes colour when it comes into contact with a defective region in a sample. The device might be used to monitor structures such as rail tracks and pipelines.

## Magnetically Responsive Photonic Crystals Nanofluids

Magnetic nanoparticle clusters or magnetic nanobeads with the size 80–150 nanometers form ordered structures along the direction of the external magnetic field with a regular interparticle spacing on the order of hundreds of nanometers resulting in strong diffraction of visible light in suspension.

## Applications

Nanofluids are primarily used for their enhanced thermal properties as coolants in heat transfer equipment such as heat exchangers, electronic cooling system(such as flat plate) and radiators. Heat transfer over flat plate has been analyzed by many researchers. However, they are also useful for their controlled optical properties. Graphene based nanofluid has been found to enhance Polymerase chain reaction efficiency. Nanofluids in solar collectors is another application where nanofluids are employed for their tunable optical properties.

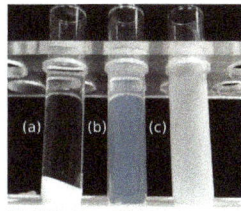

## Notable Researchers

- Stephen U. S. Choi

- Sarit Kumar Das

- Slavko Kralj

- Thalappil Pradeep

- Jacopo Buongiorno

- Dhananjay Yadav

- Robert A Taylor

- Aminreza Noghrehabadi

- Amir Malvandi

# Nanomaterials

Nanomaterials describe, in principle, materials of which a single unit is sized (in at least one dimension) between 1 and 1000 nanometres ($10^{-9}$ meter) but is usually 1—100 nm (the usual definition of nanoscale).

Nanomaterials research takes a materials science-based approach to nanotechnology, leveraging advances in materials metrology and synthesis which have been developed in support of microfabrication research. Materials with structure at the nanoscale often have unique optical, electronic, or mechanical properties.

Nanomaterials are slowly becoming commercialized and beginning to emerge as commodities.

## Types

### Natural Nanomaterials

Biological systems often feature natural, functional nanomaterials. The structure of foraminifera (mainly chalk) and viruses (protein, capsid), the wax crystals covering a lotus or nasturtium leaf, spider and spider-mite silk, the blue hue of tarantulas, the "spatulae" on the bottom of gecko feet, some butterfly wing scales, natural colloids (milk, blood), horny materials (skin, claws, beaks, feathers, horns, hair), paper, cotton, nacre, corals, and even our own bone matrix are all natural *organic* nanomaterials.

Natural *inorganic* nanomaterials occur through crystal growth in the diverse chemical conditions of the Earth's crust. For example, clays display complex nanostructures due to anisotropy of their underlying crystal structure, and volcanic activity can give rise to opals, which are an instance of a naturally occurring photonic crystals due to their nanoscale structure. Fires represent particularly complex reactions and can produce pigments, cement, fumed silica etc.

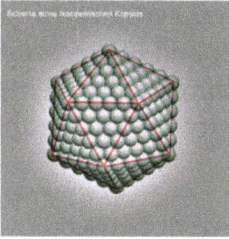

Viral capsid

"Lotus effect", hydrophobic effect with self-cleaning ability

Close-up of the underside of a gecko's foot as it walks on a glass wall. (spatula: $200 \times 10\text{-}15$ nm).

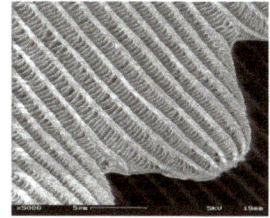

REM scan of a butterfly wing scale ($\times 5000$)

Peacock wing (detail)

Brazilian Crystal Opal. The play of color is caused by the interference and diffraction of light between silica spheres (150 - 300 nm in diameter).

Lycurgus Cup, glass, 4th century, Roman. Nanoparticles (70 nm) of gold and silver, dispersed in colloidal form, are responsible for the dichroic effect (red/green).

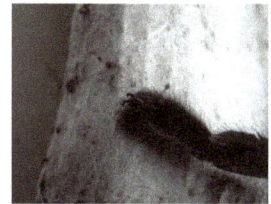

Blue hue of a species of tarantula (450 nm ± 20 nm)

## Fullerenes

The fullerenes are a class of allotropes of carbon which conceptually are graphene sheets rolled into tubes or spheres. These include the carbon nanotubes (or silicon nanotubes) which are of interest both because of their mechanical strength and also because of their electrical properties.

The first fullerene molecule to be discovered, and the family's namesake, buckminsterfullerene ($C_{60}$), was prepared in 1985 by Richard Smalley, Robert Curl, James Heath, Sean O'Brien, and Harold Kroto at Rice University. The name was a homage to Buckminster Fuller, whose geodesic domes it resembles. Fullerenes have since been found to occur in nature. More recently, fullerenes have been detected in outer space.

For the past decade, the chemical and physical properties of fullerenes have been a hot topic in the field of research and development, and are likely to continue to be for a long time. In April 2003, fullerenes were under study for potential medicinal use: binding specific antibiotics to the structure of resistant bacteria and even target certain types of cancer cells such as melanoma. The October 2005 issue of Chemistry and Biology contains an article describing the use of fullerenes as light-activated antimicrobial agents. In the field of nanotechnology, heat resistance and superconductivity are among the properties attracting intense research.

A common method used to produce fullerenes is to send a large current between two nearby graphite electrodes in an inert atmosphere. The resulting carbon plasma arc between the electrodes cools into sooty residue from which many fullerenes can be isolated.

There are many calculations that have been done using ab-initio Quantum Methods applied to fullerenes. By DFT and TDDFT methods one can obtain IR, Raman and UV spectra. Results of such calculations can be compared with experimental results.

## Graphene Nanostructures

2D materials are crystalline materials consisting of a two-dimensional single layer of atoms. The most important representative graphene was discovered in 2004. Other 2D materials based on other elements have since been reported.

Box-shaped graphene (BSG) nanostructure is an example of 3D nanomaterial. BSG nanostructure has appeared after mechanical cleavage of pyrolytic graphite. This nanostructure is a multilayer system of parallel hollow nanochannels located along the surface and having quadrangular cross-section. The thickness of the channel walls is approximately equal to 1 nm. The typical width of channel facets makes about 25 nm.

## Nanoparticles

Inorganic nanomaterials, (e.g. quantum dots, nanowires and nanorods) because of their interesting optical and electrical properties, could be used in optoelectronics. Furthermore, the optical and electronic properties of nanomaterials which depend on their size and shape can be tuned via synthetic techniques. There are the possibilities to use those materials in organic material based optoelectronic devices such as Organic solar cells, OLEDs etc. The operating principles of such devices are governed by photoinduced processes like electron transfer and energy transfer. The performance of the devices depends on the efficiency of the photoinduced process responsible for their functioning. Therefore, better understanding of those photoinduced processes in organic/inorganic nanomaterial composite systems is necessary in order to use them in organic optoelectronic devices.

Nanoparticles or nanocrystals made of metals, semiconductors, or oxides are of particular interest for their mechanical, electrical, magnetic, optical, chemical and other properties. Nanoparticles have been used as quantum dots and as chemical catalysts such as nanomaterial-based catalysts. Recently, a range of nanoparticles are extensively investigated for biomedical applications including tissue engineering, drug delivery, biosensor.

Nanoparticles are of great scientific interest as they are effectively a bridge between bulk materials and atomic or molecular structures. A bulk material should have constant physical properties regardless of its size, but at the nano-scale this is often not the case. Size-dependent properties are observed such as quantum confinement in semiconductor particles, surface plasmon resonance in some metal particles and superparamagnetism in magnetic materials.

Nanoparticles exhibit a number of special properties relative to bulk material. For example, the bending of bulk copper (wire, ribbon, etc.) occurs with movement of copper atoms/clusters at about the 50 nm scale. Copper nanoparticles smaller than 50 nm are considered super hard materials that do not exhibit the same malleability and ductility as bulk copper. The change in properties is not always desirable. Ferroelectric materials smaller than 10 nm can switch their magnetisation direction using room temperature thermal energy, thus making them useless for memory storage. Suspensions of nanoparticles are possible because the interaction of the particle surface with the solvent is strong enough to overcome differences in density, which usually result in a material either sinking or floating in a liquid. Nanoparticles often have unexpected visual properties because they are small enough to confine their electrons and produce quantum effects. For example, gold nanoparticles appear deep red to black in solution.

The often very high surface area to volume ratio of nanoparticles provides a tremendous driving force for diffusion, especially at elevated temperatures. Sintering is possible at lower temperatures and over shorter durations than for larger particles. This theoretically does not affect the density of the final product, though flow difficulties and the tendency of nanoparticles to agglomerate do complicate matters. The surface effects of nanoparticles also reduces the incipient melting temperature.

## Nanozymes

Nanozymes are nanomaterials with enzyme-like characteristics. They are an emerging type of artificial enzyme, which have been used for wide applications in such as biosensing, bioimaging, tumor diagnosis and therapy, antibiofouling, etc.

## Synthesis

The goal of any synthetic method for nanomaterials is to yield a material that exhibits properties that are a result of their characteristic length scale being in the nanometer range (~1 – 100 nm). Accordingly, the synthetic method should exhibit control of size in this range so that one property or another can be attained. Often the methods are divided into two main types "Bottom Up" and "Top Down."

## Bottom up Methods

Bottom up methods involve the assembly of atoms or molecules into nanostructured arrays. In these methods the raw material sources can be in the form of gases, liquids or solids. The latter requiring some sort of disassembly prior to their incorporation onto a nanostructure. Bottom methods generally fall into two categories: chaotic and controlled.

## Chaotic Processes

Chaotic processes involve elevating the constituent atoms or molecules to a chaotic state and then suddenly changing the conditions so as to make that state unstable. Through the clever manipulation of any number of parameters, products form largely as a result of the insuring kinetics. The collapse from the chaotic state can be difficult or impossible to control and so ensemble statistics often govern the resulting size distribution and average size. Accordingly, nanoparticle formation is controlled through manipulation of the end state of the products.

Examples of Chaotic Processes are: Laser ablation, Exploding wire, Arc, Flame pyrolysis, Combustion, Precipitation synthesis techniques.

## Controlled Processes

Controlled Processes involve the controlled delivery of the constituent atoms or molecules to the site(s) of nanoparticle formation such that the nanoparticle can grow to a prescribed sizes in a controlled manner. Generally the state of the constituent atoms or molecules are never far from that needed for nanoparticle formation. Accordingly, nanoparticle formation is controlled through the control of the state of the reactants.

Examples of controlled processes are self-limiting growth solution, self-limited chemical vapor deposition, shaped pulse femtosecond laser techniques, and molecular beam epitaxy.

## Characterization

Novel effects can occur in materials when structures are formed with sizes comparable to any one of many possible length scales, such as the de Broglie wavelength of electrons, or the optical wavelengths of high energy photons. In these cases quantum mechanical effects can dominate material properties. One example is quantum confinement where the electronic properties of solids are altered with great reductions in particle size. The optical properties of nanoparticles, e.g. fluorescence, also become a function of the particle diameter. This effect does not come into play by going from macrosocopic to micrometer dimensions, but becomes pronounced when the nanometer scale is reached.

In addition to optical and electronic properties, the novel mechanical properties of many nanomaterials is the subject of nanomechanics research. When added to a bulk material, nanoparticles can strongly influence the mechanical properties of the material, such as the stiffness or elasticity. For example, traditional polymers can be reinforced by nanoparticles (such as carbon nanotubes) resulting in novel materials which can be used as lightweight replacements for metals. Such composite materials may enable a weight reduction accompanied by an increase in stability and improved functionality.

Finally, nanostructured materials with small particle size such as zeolites, and asbestos, are used as catalysts in a wide range of critical industrial chemical reactions. The further development of such catalysts can form the basis of more efficient, environmentally friendly chemical processes.

The first observations and size measurements of nano-particles were made during the first decade of the 20th century. Zsigmondy made detailed studies of gold sols and other nanomaterials with sizes down to 10 nm and less. He published a book in 1914. He used an ultramicroscope that employs a *dark field* method for seeing particles with sizes much less than light wavelength.

There are traditional techniques developed during 20th century in Interface and Colloid Science for characterizing nanomaterials. These are widely used for *first generation* passive nanomaterials specified in the next section.

These methods include several different techniques for characterizing particle size distribution. This characterization is imperative because many materials that are expected to be nano-sized are actually aggregated in solutions. Some of methods are based on light scattering. Others apply ultrasound, such as ultrasound attenuation spectroscopy for testing concentrated nano-dispersions and microemulsions.

There is also a group of traditional techniques for characterizing surface charge or zeta potential of nano-particles in solutions. This information is required for proper system stabilzation, preventing its aggregation or flocculation. These methods include micro-electrophoresis, electrophoretic light scattering and electroacoustics. The last one, for instance colloid vibration current method is suitable for characterizing concentrated systems.

## Uniformity

The chemical processing and synthesis of high performance technological components for the private, industrial and military sectors requires the use of high purity ceramics, polymers, glass-ceramics and material composites. In condensed bodies formed from fine powders, the irregular sizes and shapes of nanoparticles in a typical powder often lead to non-uniform packing morphologies that result in packing density variations in the powder compact.

Uncontrolled agglomeration of powders due to attractive van der Waals forces can also give rise to in microstructural inhomogeneities. Differential stresses that develop as a result of non-uniform drying shrinkage are directly related to the rate at which the solvent can be removed, and thus highly dependent upon the distribution of porosity. Such stresses have been associated with a plastic-to-brittle transition in consolidated bodies, and can yield to crack propagation in the unfired body if not relieved.

In addition, any fluctuations in packing density in the compact as it is prepared for the kiln are often amplified during the sintering process, yielding inhomogeneous densification. Some pores and other structural defects associated with density variations have been shown to play a detrimental role in the sintering process by growing and thus limiting end-point densities. Differential stresses arising from inhomogeneous densification have also been shown to result in the propagation of internal cracks, thus becoming the strength-controlling flaws.

It would therefore appear desirable to process a material in such a way that it is physically uniform with regard to the distribution of components and porosity, rather than using particle size distributions which will maximize the green density. The containment of a uniformly dispersed assembly of strongly interacting particles in suspension requires total control over particle-particle interactions. It should be noted here that a number of dispersants such as ammonium citrate (aqueous) and imidazoline or oleyl alcohol (nonaqueous) are promising solutions as possible additives for enhanced dispersion and deagglomeration. Monodisperse nanoparticles and colloids provide this potential.

Monodisperse powders of colloidal silica, for example, may therefore be stabilized sufficiently to ensure a high degree of order in the colloidal crystal or polycrystalline colloidal solid which results from aggregation. The degree of order appears to be limited by the time and space allowed for longer-range correlations to be established. Such defective polycrystalline colloidal structures would appear to be the basic elements of sub-micrometer colloidal materials science, and, therefore, provide the first step in developing a more rigorous understanding of the mechanisms involved in microstructural evolution in high performance materials and components.

## Legal Definition

On 18 October 2011, the European Commission adopted the following definition of a nanomaterial: "A natural, incidental or manufactured material containing particles, in an unbound state or as an aggregate or as an agglomerate and where, for 50% or more of the particles in the number size distribution, one or more external dimensions is in the size range 1 nm – 100 nm. In specific cases and where warranted by concerns for the environment, health, safety or competitiveness the number size distribution threshold of 50% may be replaced by a threshold between 1 and 50%."

However, this differs from the definition adopted by the International Organization for Standardization (ISO), which is: "Material with any external dimension in the nanoscale or having internal structure in the nanoscale."

"Nanoscale" is, in turn, defined as: "Size range from approximately 1 nm to 100 nm."

It is not currently known which of these, if any, will prevail in courts of law.

## Safety

## Health Impact

While nanomaterials and nanotechnologies are expected to yield numerous health and health care advances, such as more targeted methods of delivering drugs, new cancer therapies, and methods of early detection of diseases, they also may have unwanted effects. Increased rate of absorption is the main concern associated with manufactured nanoparticles.

When materials are made into nanoparticles, their surface area to volume ratio increases. The greater specific surface area (surface area per unit weight) may lead to increased rate of absorption through the skin, lungs, or digestive tract and may cause unwanted effects to the lungs as well as other organs. However, the particles must be absorbed in sufficient quantities in order to pose health risks.

Nanoparticles created adventitiously (e.g., through the rubbing of prostheses) have long been known to be a health hazard, but as the use of nanomaterials increases worldwide, concerns for worker and user safety are mounting. To address such concerns, the Swedish Karolinska Institute conducted a study in which various nanoparticles were introduced to human lung epithelial cells. The results, released in 2008, showed that iron oxide nanoparticles caused little DNA damage and were non-toxic. Zinc oxide nanoparticles were slightly worse. Titanium dioxide caused only DNA damage. Carbon nanotubes caused DNA damage at low levels. Copper oxide was found to be the worst offender, and was the only nanomaterial identified by the researchers as a clear health risk. Though nanomaterials are not confirmed as a health risk to workers who produce them, NIOSH recommends that exposure precautions and personal protective equipment be used to protect workers until the risks of nanomaterial manufacture are better understood.

## Occupational Safety

Beyond normal chemical safety practices such as using personal protective equipment, The United States National Institute for Occupational Safety and Health recommends the use of HEPA filtration on local exhaust ventilation such as fume hoods. General laboratory ventilation is often not sufficient to effectively clear nanomaterials released into the general room air over a 30-minute period; therefore, researchers should leave the hood fan on even after synthesis is complete. Nanomaterials in dry powder form have the most risk of airborne dust generation and dermal contact, followed by liquid suspensions and then by nanomaterials embedded in a solid matrix.

Nanoparticles behave differently than other similarly sized particles. It is therefore necessary to develop specialized approaches to testing and monitoring their effects on human health and on the environment. The OECD Chemicals Committee has established the Working Party on Manufactured Nanomaterials to address this issue and to study the practices of OECD member countries in regards to nanomaterial safety.

## Nanoparticle

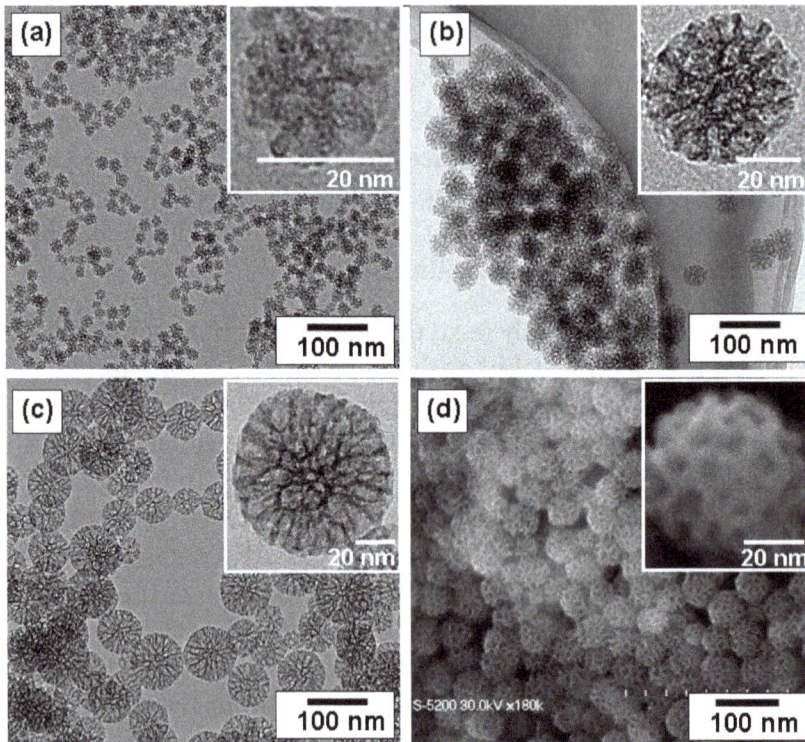

TEM (a, b, and c) images of prepared mesoporous silica nanoparticles with mean outer diameter: (a) 20nm, (b) 45nm, and (c) 80nm. SEM (d) image corresponding to (b). The insets are a high magnification of mesoporous silica particle.

Nanoparticles are particles between 1 and 100 nanometers in size. In nanotechnology, a particle is defined as a small object that behaves as a whole unit with respect to its transport and properties. Particles are further classified according to diameter. Ultrafine particles are the same as nanoparticles and between 1 and 100 nanometers in size, fine particles are sized between 100 and 2,500 nanometers, and coarse particles cover a range between 2,500 and 10,000 nanometers. Scientific research on nanoparticles is intense as they have many potential applications in medicine, optics, and electronics. The U.S. National Nanotechnology Initiative has offers government funding focused on nanoparticle research.

## Definition

### IUPAC Definition

Particle of any shape with dimensions in the $1 \times 10^{-9}$ and $1 \times 10^{-7}$ m range.

The term nanoparticle is not usually applied to individual molecules, and usually refers to inorganic materials.

The reason for the synonymous definition of nanoparticles and ultrafine particles is that, during the 1970-80s, when the first thorough fundamental studies with "nanoparticles" were underway in the USA (by Granqvist and Buhrman) and Japan, (within an ERATO Project) they were called "ultrafine particles" (UFP). However, during the 1990s before the National Nanotechnology Initiative was launched in the USA, the new name, "nanoparticle," had become fashionable. Nanoparticles can exhibit size-related properties significantly different from those of either fine particles or bulk materials.

Nanoclusters have at least one dimension between 1 and 10 nanometers and a narrow size distribution. Nanopowders are agglomerates of ultrafine particles, nanoparticles, or nanoclusters. Nanometer-sized single crystals, or single-domain ultrafine particles, are often referred to as nanocrystals.

## Background

Although nanoparticles are associated with modern science, they have a long history. Nanoparticles were used by artisans as far back as the ninth century in Mesopotamia for generating a glittering effect on the surface of pots.

Even these days, pottery from the Middle Ages and Renaissance often retain a distinct gold- or copper-colored metallic glitter. This luster is caused by a metallic film that was applied to the transparent surface of a glazing. The luster can still be visible if the film has resisted atmospheric oxidation and other weathering.

The luster originated within the film itself, which contained silver and copper nanoparticles dispersed homogeneously in the glassy matrix of the ceramic glaze. These nanoparticles were created by the artisans by adding copper and silver salts and oxides together with vinegar, ochre, and clay on the surface of previously-glazed pottery. The object was then placed into a kiln and heated to about 600 °C in a reducing atmosphere.

In the heat the glaze would soften, causing the copper and silver ions to migrate into the outer layers of the glaze. There the reducing atmosphere reduced the ions back to metals, which then came together forming the nanoparticles that give the colour and optical effects.

Luster technique showed that ancient craftsmen had a rather sophisticated empirical knowledge of materials. The technique originated in the Muslim world. As Muslims were not allowed to use gold in artistic representations, they sought a way to create a similar effect without using real gold. The solution they found was using luster.

Michael Faraday provided the first description, in scientific terms, of the optical properties of nanometer-scale metals in his classic 1857 paper. In a subsequent paper, the author (Turner) points out that: "It is well known that when thin leaves of gold or silver are mounted upon glass and heated to a temperature that is well below a red heat (~500 °C), a remarkable change of properties takes place, whereby the continuity of the metallic film is destroyed. The result is that white light is now freely transmitted, reflection is correspondingly diminished, while the electrical resistivity is enormously increased."

## Uniformity

The chemical processing and synthesis of high-performance technological components for the private, industrial, and military sectors requires the use of high-purity ceramics, polymers, glass-ceramics, and composite materials. In condensed bodies formed from fine powders, the irregular particle sizes and shapes in a typical powder often lead to non-uniform packing morphologies that result in packing density variations in the powder compact.

Uncontrolled agglomeration of powders due to attractive van der Waals forces can also give rise to in microstructural heterogeneity. Differential stresses that develop as a result of non-uniform drying shrinkage are directly related to the rate at which the solvent can be removed, and thus highly dependent upon the distribution of porosity. Such stresses have been associated with a plastic-to-brittle transition in consolidated bodies, and can yield to crack propagation in the unfired body if not relieved.

In addition, any fluctuations in packing density in the compact as it is prepared for the kiln are often amplified during the sintering process, yielding inhomogeneous densification. Some pores and other structural defects associated with density variations have been shown to play a detrimental role in the sintering process by growing and thus limiting end-point densities. Differential stresses arising from inhomogeneous densification have also been shown to result in the propagation of internal cracks, thus becoming the strength-controlling flaws.

Inert gas evaporation and inert gas deposition are free many of these defects due to the distillation (cf. purification) nature of the process and having enough time to form single crystal particles, however even their non-aggreated deposits have lognormal size distribution, which is typical with nanoparticles. The reason why modern gas evaporation techniques can produce a relatively narrow size distribution is that aggregation can be avoided. However, even in this case, random residence times in the growth zone, due to the combination of drift and diffusion, result in a size distribution appearing lognormal.

It would, therefore, appear desirable to process a material in such a way that it is physically uniform with regard to the distribution of components and porosity, rather than using particle size distributions that will maximize the green density. The containment of a uniformly dispersed assembly of strongly interacting particles in suspension requires total control over interparticle forces. Monodisperse nanoparticles and colloids provide this potential.

Monodisperse powders of colloidal silica, for example, may therefore be stabilized sufficiently to ensure a high degree of order in the colloidal crystal or polycrystalline colloidal solid that results from aggregation. The degree of order appears to be limited by the time and space allowed for longer-range correlations to be established. Such defective polycrystalline colloidal structures would appear to be the basic elements of submicrometer colloidal materials science and, therefore, provide the first step in developing a more rigorous understanding of the mechanisms involved in microstructural evolution in high performance materials and components.

## Properties

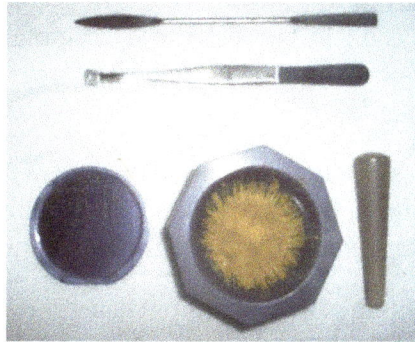

Silicon nanopowder

1 kg of particles of 1 mm³ has the same surface area as 1 mg of particles of 1 nm³

Nanoparticles are of great scientific interest as they are, in effect, a bridge between bulk materials and atomic or molecular structures. A bulk material should have constant physical properties regardless of its size, but at the nano-scale size-dependent properties are often observed. Thus, the properties of materials change as their size approaches the nanoscale and as the percentage of the surface in relation to the percentage of the volume of a material becomes significant. For bulk materials larger than one micrometer (or micron), the percentage of the surface is insignificant in relation to the volume in the bulk of the material. *The interesting and sometimes unexpected properties of nanoparticles are therefore largely due to the large surface area of the material, which dominates the contributions made by the small bulk of the material.*

Nanoparticles often possess unexpected optical properties as they are small enough to confine their electrons and produce quantum effects. For example, gold nanoparticles appear deep-red to black in solution. Nanoparticles of yellow gold and grey silicon are red in color. Gold nanoparticles melt at much lower temperatures (~300 °C for 2.5 nm size) than the gold slabs (1064 °C);. Absorption of solar radiation is much higher in materials composed of nanoparticles than it is in thin films of continuous sheets of material. In both solar PV and solar thermal applications, controlling the size, shape, and material of the particles, it is possible to control solar absorption.

Other size-dependent property changes include quantum confinement in semiconductor particles, surface plasmon resonance in some metal particles and superparamagnetism in magnetic materials. What would appear ironic is that the changes in physical properties are not always desirable. Ferromagnetic materials smaller than 10 nm can switch their magnetisation direction using room temperature thermal energy, thus making them unsuitable for memory storage.

Suspensions of nanoparticles are possible since the interaction of the particle surface with the solvent is strong enough to overcome density differences, which otherwise usually result in a material either sinking or floating in a liquid.

The high surface area to volume ratio of nanoparticles provides a tremendous driving force for diffusion, especially at elevated temperatures. Sintering can take place at lower temperatures, over shorter time scales than for larger particles. In theory, this does not affect the density of the final product, though flow difficulties and the tendency of nanoparticles to agglomerate complicates matters. Moreover, nanoparticles have been found to impart some extra properties to various day to day products. For example, the presence of titanium dioxide nanoparticles imparts what we call the self-cleaning effect, and, the size being nano-range, the particles cannot be observed. Zinc oxide particles have been found to have superior UV blocking properties compared to its bulk substitute. This is one of the reasons why it is often used in the preparation of sunscreen lotions, is completely photostable and toxic.

Clay nanoparticles when incorporated into polymer matrices increase reinforcement, leading to stronger plastics, verifiable by a higher glass transition temperature and other mechanical property tests. These nanoparticles are hard, and impart their properties to the polymer (plastic). Nanoparticles have also been attached to textile fibers in order to create smart and functional clothing.

Metal, dielectric, and semiconductor nanoparticles have been formed, as well as hybrid structures (e.g., core–shell nanoparticles). Nanoparticles made of semiconducting material may also be labeled quantum dots if they are small enough (typically sub 10 nm) that quantization of electronic energy levels occurs. Such nanoscale particles are used in biomedical applications as drug carriers or imaging agents.

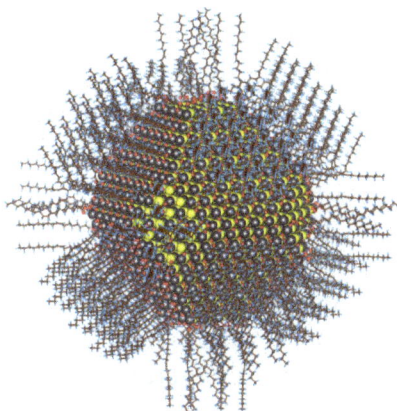

Semiconductor nanoparticle (quantum dot) of lead sulfide with complete passivation by oleic acid, oleyl and hydroxyl (size ~5nm)

Semi-solid and soft nanoparticles have been manufactured. A prototype nanoparticle of semi-solid nature is the liposome. Various types of liposome nanoparticles are currently used clinically as delivery systems for anticancer drugs and vaccines.

Nanoparticles with one half hydrophilic and the other half hydrophobic are termed Janus particles and are particularly effective for stabilizing emulsions. They can self-assemble at water/oil interfaces and act as solid surfactants.

## Synthesis

There are several methods for creating nanoparticles, including attrition, pyrolysis and hydrothermal synthesis. In attrition, macro- or micro-scale particles are ground in a ball mill, a planetary ball mill, or other size-reducing mechanism. The resulting particles are air classified to recover nanoparticles. In pyrolysis, a vaporous precursor (liquid or gas) is forced through an orifice at high pressure and burned. The resulting solid (a version of soot) is air classified to recover oxide particles from by-product gases. Traditional pyrolysis often results in aggregates and agglomerates rather than single primary particles. Ultrasonic nozzle spray pyrolysis (USP) on the other hand aids in preventing agglomerates from forming.

A thermal plasma can deliver the energy to vaporize small micrometer-size particles. The thermal plasma temperatures are in the order of 10,000 K, so that solid powder easily evaporates. Nanoparticles are formed upon cooling while exiting the plasma region. The main types of the thermal plasma torches used to produce nanoparticles are dc plasma jet, dc arc plasma, and radio frequency (RF) induction plasmas. In the arc plasma reactors, the energy necessary for evaporation and reaction is provided by an electric arc formed between the anode and the cathode. For example, silica sand can be vaporized with an arc plasma at atmospheric pressure, or thin aluminum wires can be vaporized by exploding wire method. The resulting mixture of plasma gas and silica vapour can be rapidly cooled by quenching with oxygen, thus ensuring the quality of the fumed silica produced.

In RF induction plasma torches, energy coupling to the plasma is accomplished through the electromagnetic field generated by the induction coil. The plasma gas does not come in contact with electrodes, thus eliminating possible sources of contamination and allowing the operation of such plasma torches with a wide range of gases including inert, reducing, oxidizing, and other corrosive atmospheres. The working frequency is typically between 200 kHz and 40 MHz. Laboratory units run at power levels in the order of 30–50 kW, whereas the large-scale industrial units have been tested at power levels up to 1 MW. As the residence time of the injected feed droplets in the plasma is very short, it is important that the droplet sizes are small enough in order to obtain complete evaporation. The RF plasma method has been used to synthesize different nanoparticle materials, for example synthesis of various ceramic nanoparticles such as oxides, carbours/carbides, and nitrides of Ti and Si.

Inert-gas condensation is frequently used to make nanoparticles from metals with low melting points. The metal is vaporized in a vacuum chamber and then supercooled with an inert gas stream. The supercooled metal vapor condenses into nanometer-size particles, which can be entrained in the inert gas stream and deposited on a substrate or studied in situ.

Nanoparticles can also be formed using radiation chemistry. Radiolysis from gamma rays can create strongly active free radicals in solution. This relatively simple technique uses a minimum number of chemicals. These including water, a soluble metallic salt, a radical scavenger (often a secondary alcohol), and a surfactant (organic capping agent). High gamma doses on the order of $10^4$ Gray are required. In this process, reducing radicals will drop metallic ions down to the zero-valence state. A scavenger chemical will preferentially interact with oxidizing radicals to prevent the re-oxidation of the metal. Once in the zero-valence state, metal atoms begin to coalesce into particles. A chemical surfactant surrounds the particle during formation and regulates its growth. In sufficient concentrations, the surfactant molecules stay attached to the particle. This prevents it from dissociating or forming clusters with other particles. Formation of nanoparticles using the radiolysis method allows for tailoring of particle size and shape by adjusting precursor concentrations and gamma dose.

# Sol-gel

The sol-gel process is a wet-chemical technique (also known as chemical solution deposition) widely used recently in the fields of materials science and ceramic engineering. Such methods are used primarily for the fabrication of materials (typically a metal oxide) starting from a chemical solution (*sol*, short for solution), which acts as the precursor for an integrated network (or *gel*) of either discrete particles or network polymers.

Typical precursors are metal alkoxides and metal chlorides, which undergo hydrolysis and polycondensation reactions to form either a network "elastic solid" or a colloidal suspension (or dispersion) – a system composed of discrete (often amorphous) submicrometer particles dispersed to various degrees in a host fluid. Formation of a metal oxide involves connecting the metal centers with oxo (M-O-M) or hydroxo (M-OH-M) bridges, therefore generating metal-oxo or metal-hydroxo polymers in solution. Thus, the sol evolves toward the formation of a gel-like diphasic system containing both a liquid phase and solid phase whose morphologies range from discrete particles to continuous polymer networks.

In the case of the colloid, the volume fraction of particles (or particle density) may be so low that a significant amount of fluid may need to be removed initially for the gel-like properties to be recognized. This can be accomplished in a number of ways. The most simple method is to allow time for sedimentation to occur, and then pour off the remaining liquid. Centrifugation can also be used to accelerate the process of phase separation.

Removal of the remaining liquid (solvent) phase requires a drying process, which is typically causes shrinkage and densification. The rate at which the solvent can be removed is ultimately determined by the distribution of porosity in the gel. The ultimate microstructure of the final component will clearly be strongly influenced by changes implemented during this phase of processing. Afterward, a thermal treatment, or firing process, is often necessary in order to favor further polycondensation and enhance mechanical properties and structural stability via final sintering, densification, and grain growth. One of the distinct advantages of using this methodology as opposed to the more traditional processing techniques is that densification is often achieved at a much lower temperature.

The precursor sol can be either deposited on a substrate to form a film (e.g., by dip-coating or spin-coating), cast into a suitable container with the desired shape (e.g., to obtain a monolithic ceramics, glasses, fibers, membranes, aerogels), or used to synthesize powders (e.g., microspheres, nanospheres). The sol-gel approach is a cheap and low-temperature technique that allows for the fine control of the product's chemical composition. Even small quantities of dopants, such as organic dyes and rare earth metals, can be introduced in the sol and end up uniformly dispersed in the final product. It can be used in ceramics processing and manufacturing as an investment casting material, or as a means of producing very thin films of metal oxides for various purposes. Sol-gel derived materials have diverse applications in optics, electronics, energy, space, (bio)sensors, medicine (e.g., controlled drug release) and separation (e.g., chromatography) technology.

## Colloids

The term colloid is used primarily to describe a range of mixtures which have solid particles dispersed in a liquid medium. The term applies only if the particles are larger than atomic dimensions but small enough to exhibit Brownian motion. If the particles are large enough, their dynamic behavior in any given period of time in suspension would be governed by forces of gravity and sedimentation. If the particles are small enough, their irregular motion in suspension can be attributed to the collective bombardment of a myriad of thermally agitated molecules in the liquid suspending medium, as described originally by Albert Einstein in his dissertation. Einstein showed evidence that water was made up of discrete molecules by characterizing the observed erratic particle behavior using the theory of Brownian motion, with sedimentation being a possible result. The critical size range (or particle diameter) typically ranges from nanometers ($10^{-9}$ m) to micrometers ($10^{-6}$ m).

## Morphology

Nanostars of vanadium(IV) oxide

Scientists have taken to naming their particles after the real-world shapes that they might represent. Nanospheres, nanochains, nanoreefs, nanoboxes and more have appeared in the literature. These morphologies sometimes arise spontaneously as an effect of a templating or directing agent present in the synthesis such as miscellar emulsions or anodized alumina pores, or from the innate crystallographic growth patterns of the materials themselves. Some of these morphologies may serve a purpose, such as long carbon nanotubes used to bridge an electrical junction, or just a scientific curiosity like the stars shown at right.

Amorphous particles usually adopt a spherical shape (due to their microstructural isotropy), whereas the shape of anisotropic microcrystalline whiskers corresponds to their particular crystal habit. At the small end of the size range, nanoparticles are often referred to as clusters. Spheres, rods, fibers, and cups are just a few of the shapes that have been grown. The study of fine particles is called micromeritics.

## Characterization

The majority of nanoparticle characterization techniques are light-based, but a non-optical nanoparticle characterization technique called Tunable Resistive Pulse Sensing (TRPS) has been developed that enables the simultaneous measurement of size, concentration and surface charge for a wide variety of nanoparticles. This technique, which applies the Coulter Principle, allows for particle-by-particle quantification of these three nanoparticle characteristics with high resolution.

## Functionalization

The surface coating of nanoparticles determines many of their properties, notably stability, solubility, and targeting. A coating that is multivalent or polymeric confers high stability. Functionalized nanomaterial-based catalysts can be used for catalysis of many known organic reactions.

## Surface Coating for Biological Applications

For biological applications, the surface coating should be polar to give high aqueous solubility and prevent nanoparticle aggregation. In serum or on the cell surface, highly charged coatings promote non-specific binding, whereas polyethylene glycol linked to terminal hydroxyl or methoxy groups repel non-specific interactions. Nanoparticles can be linked to biological molecules that can act as address tags, to direct the nanoparticles to specific sites within the body, specific organelles within the cell, or to follow specifically the movement of individual protein or RNA molecules in living cells. Common address tags are monoclonal antibodies, aptamers, streptavidin or peptides. These targeting agents should ideally be covalently linked to the nanoparticle and should be present in a controlled number per nanoparticle. Multivalent nanoparticles, bearing multiple targeting groups, can cluster receptors, which can activate cellular signaling pathways, and give stronger anchoring. Monovalent nanoparticles, bearing a single binding site, avoid clustering and so are preferable for tracking the behavior of individual proteins.

Red blood cell coatings can help nanoparticles evade the immune system.

## Safety

Nanoparticles present possible dangers, both medically and environmentally. Most of these are due to the high surface to volume ratio, which can make the particles very reactive or catalytic. They are also able to pass through cell membranes in organisms, and their interactions with biological systems are relatively unknown. However, it is unlikely the particles would enter the cell nucleus, Golgi complex, endoplasmic reticulum or other internal cellular components due to the particle size and intercellular agglomeration. A recent study looking at the effects of ZnO nanoparticles on human immune cells has found varying levels of susceptibility to cytotoxicity. There are concerns that pharmaceutical companies, seeking regulatory approval for nano-reformulations of existing medicines, are relying on safety data produced during clinical studies of the earlier, pre-reformulation version of the medicine. This could result in regulatory bodies, such as the FDA, missing new side effects that are specific to the nano-reformulation.

Whether cosmetics and sunscreens containing nanomaterials pose health risks remains largely unknown at this stage. However considerable research has demonstrated that zinc nanoparticles are not absorbed into the bloodstream in vivo.

Concern has also been raised over the health effects of respirable nanoparticles from certain combustion processes. As of 2013 the U.S. Environmental Protection Agency was investigating the safety of the following nanoparticles:

- Carbon Nanotubes: Carbon materials have a wide range of uses, ranging from composites for use in vehicles and sports equipment to integrated circuits for electronic components. The interactions between nanomaterials such as carbon nanotubes and natural organic matter strongly influence both their aggregation and deposition, which strongly affects their transport, transformation, and exposure in aquatic environments. In past research, carbon nanotubes exhibited some toxicological impacts that will be evaluated in various environmental settings in current EPA chemical safety research. EPA research will provide data, models, test methods, and best practices to discover the acute health effects of carbon nanotubes and identify methods to predict them.

- Cerium oxide: Nanoscale cerium oxide is used in electronics, biomedical supplies, energy, and fuel additives. Many applications of engineered cerium oxide nanoparticles naturally disperse themselves into the environment, which increases the risk of exposure. There is ongoing exposure to new diesel emissions using fuel additives containing $CeO_2$ nanoparticles, and the environmental and public health impacts of this new technology are unknown. EPA's chemical safety research is assessing the environmental, ecological, and health implications of nanotechnology-enabled diesel fuel additives.

- Titanium dioxide: Nano titanium dioxide is currently used in many products. Depending on the type of particle, it may be found in sunscreens, cosmetics, and paints and coatings. It is also being investigated for use in removing contaminants from drinking water.

- Nano Silver: Nano silver is being incorporated into textiles, clothing, food packaging, and other materials to eliminate bacteria. EPA and the U.S. Consumer Product Safety Commission are studying certain products to see whether they transfer nano-size silver particles in real-world scenarios. EPA is researching this topic to better understand how much nano-silver children come in contact with in their environments.

- Iron: While nano-scale iron is being investigated for many uses, including "smart fluids" for uses such as optics polishing and as a better-absorbed iron nutrient supplement, one of its more prominent current uses is to remove contamination from groundwater. This use, supported by EPA research, is being piloted at a number of sites across the country.

## Laser Applications

The use of nanoparticles in laser dye-doped poly(methyl methacrylate) (PMMA) laser gain media was demonstrated in 2003 and it has been shown to improve conversion efficiencies and to decrease laser beam divergence. Researchers attribute the reduction in beam divergence to improved dn/dT characteristics of the organic-inorganic dye-doped nanocomposite. The optimum composition reported by these researchers is 30% w/w of $SiO_2$ (~ 12 nm) in dye-doped PMMA.

## Medicinal Applications

- Liposome

- Dendrimer

- Iron oxide nanoparticles

- Nanomedicine

- Polymer-drug conjugate

- Polymeric nanoparticle

## Nanometre

The nanometre (International spelling as used by the International Bureau of Weights and Measures; SI symbol: nm) or nanometer (American spelling) is a unit of length in the metric system, equal to one billionth of a metre (0.000000001 m). The name combines the SI prefix nano- with the parent unit name metre. It can be written in scientific notation as $1 \times 10{-9}$ m, in engineering notation as 1 E–9 m, and is simply 1/1000000000 metres. One nanometre equals ten ångströms.

## Use

When used as a prefix for something other than a unit of measure (as in "nanoscience"), nano refers to nanotechnology, or phenomena typically occurring on a scale of nano-metres. The nanometre is often used to express dimensions on an atomic scale: the diameter of a helium atom, for example, is about 0.1 nm, and that of a ribosome is about 20 nm. The nanometre is also commonly used to specify the wavelength of elect-romagnetic radiation near the visible part of the spectrum: visible light ranges from around 400 to 700 nm. The ångström, which is equal to 0.1 nm, was formerly used for these purposes, but is still used in other fields. Since the late 1980s, in usages such as 32 nm and 22 nm, it has also been used to describe typical feature sizes in successive generations of the ITRS Roadmap for miniaturization in the semiconductor industry.

## History

The nanometre was formerly known as the millimicrometre – or, more commonly, the millimicron for short – since it is 1/1000 of a micron (micrometre), and was often de-noted by the symbol mμ or (more rarely and confusingly, since it logically should refer to a *millionth* of a micron) as μμ.

# Carbon Nanotube

Carbon nanotubes (CNTs) are allotropes of carbon with a cylindrical nanostructure. Nanotubes have been constructed with length-to-diameter ratio of up to 132,000,000:1, significantly larger than for any other material. These cylindrical carbon molecules have unusual properties, which are valuable for nanotechnology, electronics, optics and oth-er fields of materials science and technology. In particular, owing to their extraordinary thermal conductivity and mechanical and electrical properties, carbon nanotubes find applications as additives to various structural materials. For instance, nanotubes form a tiny portion of the material(s) in some (primarily carbon fiber) baseball bats, golf clubs, car parts or damascus steel.

Nanotubes are members of the fullerene structural family. Their name is derived from their long, hollow structure with the walls formed by one-atom-thick sheets of carbon, called graphene. These sheets are rolled at specific and discrete ("chiral") angles, and the combination of the rolling angle and radius decides the nanotube properties; for ex-ample, whether the individual nanotube shell is a metal or semiconductor. Nanotubes are categorized as single-walled nanotubes (SWNTs) and multi-walled nanotubes (MWNTs). Individual nanotubes naturally align themselves into "ropes" held together by van der Waals forces, more specifically, pi-stacking.

Applied quantum chemistry, specifically, orbital hybridization best describes chemi-cal bonding in nanotubes. The chemical bonding of nanotubes is composed entirely of $sp^2$ bonds, similar to those of graphite. These bonds, which are stronger than the $sp^3$ bonds found in alkanes and diamond, provide nanotubes with their unique strength.

# Types of Carbon Nanotubes and Related Structures

## Terminology

There is no consensus on some terms describing carbon nanotubes in scientific literature: both "-wall" and "-walled" are being used in combination with "single", "double",
"triple" or "multi", and the letter C is often omitted in the abbreviation; for example,
multi-walled carbon nanotube (MWNT).

## Single-walled

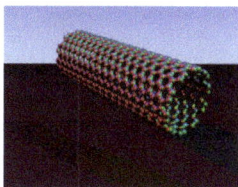 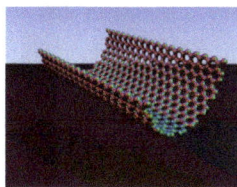 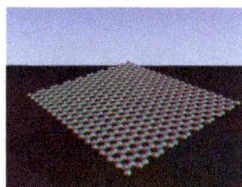 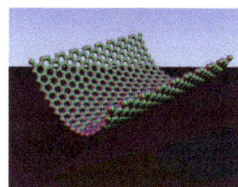

| Armchair $(n,n)$ i.e.: $m=n$ | The translation vector is bent, while the chiral vector stays straight | Graphene nanoribbon | The chiral vector is bent, while the translation vector stays straight |

 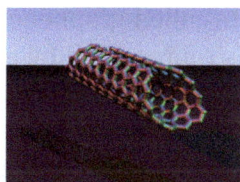 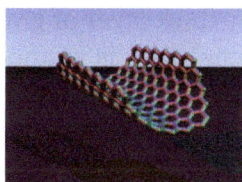 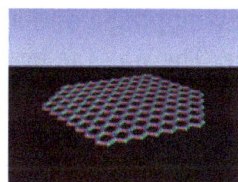

| Zigzag $(n,0)$ | Chiral $(n,m)$ | $n$ and $m$ can be counted at the end of the tube | Graphene nanoribbon |

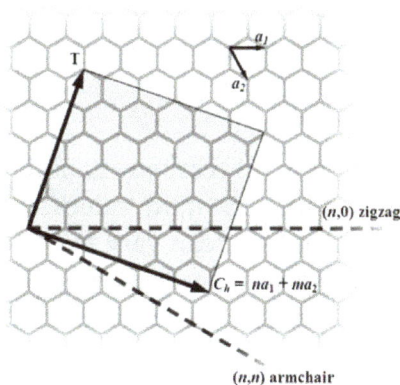

The $(n,m)$ nanotube naming scheme can be thought of as a vector ($\mathbf{C}_h$) in an infinite graphene sheet that
describes how to "roll up" the graphene sheet to make the nanotube. $\mathbf{T}$ denotes the tube axis, and $\mathbf{a}_1$ and
$\mathbf{a}_2$ are the unit vectors of graphene in real space.

A scanning tunneling microscopy image of single-walled carbon nanotube

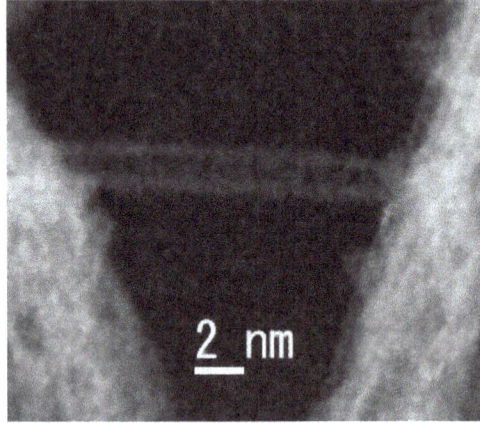

A transmission electron microscopy image of a single-walled carbon nanotube

Most single-walled nanotubes (SWNTs) have a diameter of close to 1 nanometer, and can be many millions of times longer. The structure of a SWNT can be conceptualized by wrapping a one-atom-thick layer of graphite called graphene into a seamless cylinder. The way the graphene sheet is wrapped is represented by a pair of indices $(n,m)$. The integers $n$ and $m$ denote the number of unit vectors along two directions in the honeycomb crystal lattice of graphene. If $m = 0$, the nanotubes are called zigzag nanotubes, and if $n = m$, the nanotubes are called armchair nanotubes. Otherwise, they are called chiral. The diameter of an ideal nanotube can be calculated from its (n,m) indices as follows

$$d = \frac{a}{\pi}\sqrt{(n^2 + nm + m^2)} = 78.3\sqrt{((n+m)^2 - nm)}\,\text{pm},$$

where $a = 0.246$ nm.

SWNTs are an important variety of carbon nanotube because most of their properties change significantly with the $(n,m)$ values, and this dependence is non-monotonic. In particular, their band gap can vary from zero to about 2 eV and their electrical conductivity can show metallic or semiconducting behavior. Single-walled nanotubes are likely candidates for miniaturizing electronics. The most basic building block of these systems is the electric wire, and SWNTs with diameters of an order of a nanometer can be excellent conductors. One useful application of SWNTs is in the development of the first intermolecular field-effect transistors (FET). The first intermolecular logic gate using SWCNT FETs was made in 2001. A logic gate requires both a p-FET and an n-FET. Because SWNTs are p-FETs when exposed to oxygen and n-FETs otherwise, it is possible to expose half of an SWNT to oxygen and protect the other half from it. The resulting SWNT acts as a *not* logic gate with both p and n-type FETs in the same molecule.

Prices for single-walled nanotubes declined from around $1500 per gram as of 2000 to retail prices of around $50 per gram of as-produced 40–60% by weight SWNTs as of March 2010. As of 2016 the retail price of as-produced 75% by weight SWNTs were $2 per gram, cheap enough for widespread use. SWNTs are forecast to make a large impact in electronics applications by 2020 according to the *The Global Market for Carbon Nanotubes* report.

## Multi-walled

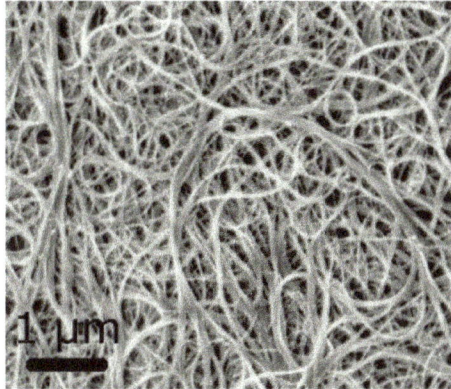

A scanning electron microscopy image of carbon nanotubes bundles

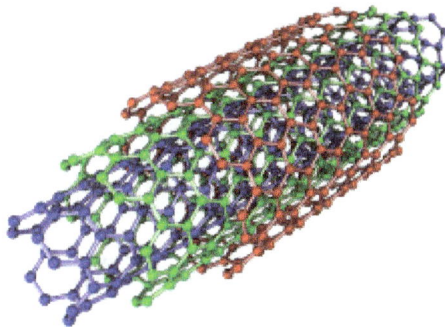

Triple-walled armchair carbon nanotube

Multi-walled nanotubes (MWNTs) consist of multiple rolled layers (concentric tubes) of graphene. There are two models that can be used to describe the structures of multi-walled nanotubes. In the *Russian Doll* model, sheets of graphite are arranged in concentric cylinders, e.g., a (0,8) single-walled nanotube (SWNT) within a larger (0,17) single-walled nanotube. In the *Parchment* model, a single sheet of graphite is rolled in around itself, resembling a scroll of parchment or a rolled newspaper. The interlayer distance in multi-walled nanotubes is close to the distance between graphene layers in graphite, approximately 3.4 Å. The Russian Doll structure is observed more commonly. Its individual shells can be described as SWNTs, which can be metallic or semiconducting. Because of statistical probability and restrictions on the relative diameters of the individual tubes, one of the shells, and thus the whole MWNT, is usually a zero-gap metal.

Double-walled carbon nanotubes (DWNTs) form a special class of nanotubes because their morphology and properties are similar to those of SWNTs but they are more resistant to chemicals. This is especially important when it is necessary to graft chemical functions to the surface of the nanotubes (functionalization) to add properties to the CNT. Covalent functionalization of SWNTs will break some C=C double bonds, leaving "holes" in the structure on the nanotube, and thus modifying both its mechanical and electrical properties. In the case of DWNTs, only the outer wall is modified. DWNT synthesis on the gram-scale was first proposed in 2003 by the CCVD technique, from the selective reduction of oxide solutions in methane and hydrogen.

The telescopic motion ability of inner shells and their unique mechanical properties will permit the use of multi-walled nanotubes as main movable arms in coming nano-mechanical devices. Retraction force that occurs to telescopic motion caused by the Lennard-Jones interaction between shells and its value is about 1.5 nN.

## Torus

In theory, a nanotorus is a carbon nanotube bent into a torus (doughnut shape). Nanotori are predicted to have many unique properties, such as magnetic moments 1000 times larger than previously expected for certain specific radii. Properties such as magnetic moment, thermal stability, etc. vary widely depending on radius of the torus and radius of the tube.

## Nanobud

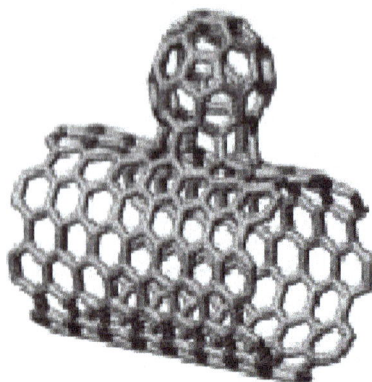

A stable nanobud structure

Carbon nanobuds are a newly created material combining two previously discovered allotropes of carbon: carbon nanotubes and fullerenes. In this new material, fullerene-like "buds" are covalently bonded to the outer sidewalls of the underlying carbon nanotube.

This hybrid material has useful properties of both fullerenes and carbon nanotubes. In particular, they have been found to be exceptionally good field emitters. In composite materials, the attached fullerene molecules may function as molecular anchors preventing slipping of the nanotubes, thus improving the composite's mechanical properties.

## Three-dimensional Carbon Nanotube Architectures

3D carbon scaffolds

Recently, several studies have highlighted the prospect of using carbon nanotubes as building blocks to fabricate three-dimensional macroscopic (>100 nm in all three dimensions) all-carbon devices. Lalwani et al. have reported a novel radical initiated thermal crosslinking method to fabricate macroscopic, free-standing, porous, all-carbon scaffolds using single- and multi-walled carbon nanotubes as building blocks. These scaffolds possess macro-, micro-, and nano- structured pores and the porosity can be tailored for specific applications. These 3D all-carbon scaffolds/architectures may be used for the fabrication of the next generation of energy storage, supercapacitors, field emission transistors, high-performance catalysis, photovoltaics, and biomedical devices and implants.

## Graphenated Carbon Nanotubes (g-CNTs)

Graphenated CNTs are a relatively new hybrid that combines graphitic foliates grown along the sidewalls of multiwalled or bamboo style CNTs. Yu *et al.* reported on "chemically bonded graphene leaves" growing along the sidewalls of CNTs. Stoner *et al.* described these structures as "graphenated CNTs" and reported in their use for enhanced supercapacitor performance. Hsu *et al.* further reported on similar structures formed on carbon fiber paper, also for use in supercapacitor applications. Pham *et al.* also reported a similar structure, namely "graphene-carbon nanotube hybrids", grown directly onto carbon fiber paper to form an integrated, binder free, high surface area conductive catalyst support for Proton Exchange Membrane Fuel Cells electrode applications with enhanced performance and durability. The foliate density can vary as a function of deposition conditions (e.g. temperature and time) with their structure ranging from few layers of graphene (< 10) to thicker, more graphite-like.

The fundamental advantage of an integrated graphene-CNT structure is the high surface area three-dimensional framework of the CNTs coupled with the high edge density of graphene. Graphene edges provide significantly higher charge density and reactivity than the basal plane, but they are difficult to arrange in a three-dimensional, high volume-density geometry. CNTs are readily aligned in a high density geometry (i.e., a vertically aligned forest) but lack high charge density surfaces—the sidewalls of the CNTs are similar to the basal plane of graphene and exhibit low charge density except where edge defects exist. Depositing a high density of graphene foliates along the length of aligned CNTs can significantly increase the total charge capacity per unit of nominal area as compared to other carbon nanostructures.

## Nitrogen-Doped Carbon Nanotubes

Nitrogen doped carbon nanotubes (N-CNTs) can be produced through five main methods, chemical vapor deposition, high-temperature and high-pressure reactions, gas-solid reaction of amorphous carbon with $NH_3$ at high temperature, solid reaction, and solvothermal synthesis.

N-CNTs can also be prepared by a CVD method of pyrolyzing melamine under Ar at elevated temperatures of 800–980 °C. However synthesis by CVD of melamine results in the formation of bamboo-structured CNTs. XPS spectra of grown N-CNTs reveal nitrogen in five main components, pyridinic nitrogen, pyrrolic nitrogen, quaternary nitrogen, and nitrogen oxides. Furthermore, synthesis temperature affects the type of nitrogen configuration.

Nitrogen doping plays a pivotal role in lithium storage, as it creates defects in the CNT walls allowing for Li ions to diffuse into interwall space. It also increases capacity by providing more favorable bind of N-doped sites. N-CNTs are also much more reactive to metal oxide nanoparticle deposition which can further enhance storage capacity, especially in anode materials for Li-ion batteries. However boron-doped nanotubes have been shown to make batteries with triple capacity.

## Peapod

A carbon peapod is a novel hybrid carbon material which traps fullerene inside a carbon nanotube. It can possess interesting magnetic properties with heating and irradiation. It can also be applied as an oscillator during theoretical investigations and predictions.

## Cup-Stacked Carbon Nanotubes

Cup-stacked carbon nanotubes (CSCNTs) differ from other quasi-1D carbon structures, which normally behave as quasi-metallic conductors of electrons. CSCNTs exhibit semiconducting behaviors due to the stacking microstructure of graphene layers.

## Extreme Carbon Nanotubes

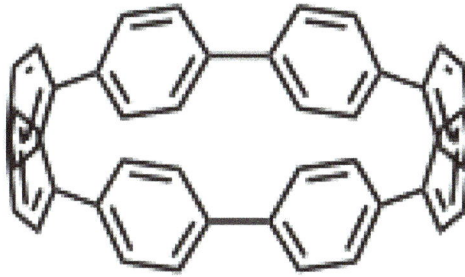

Cycloparaphenylene

The observation of the *longest* carbon nanotubes grown so far are over 1/2 m (550 mm long) was reported in 2013. These nanotubes were grown on Si substrates using an improved chemical vapor deposition (CVD) method and represent electrically uniform arrays of single-walled carbon nanotubes.

The *shortest* carbon nanotube is the organic compound cycloparaphenylene, which was synthesized in early 2009.

The *thinnest* carbon nanotube is the armchair (2,2) CNT with a diameter of 0.3 nm. This nanotube was grown inside a multi-walled carbon nanotube. Assigning of carbon nanotube type was done by a combination of high-resolution transmission electron microscopy (HRTEM), Raman spectroscopy and density functional theory (DFT) calculations.

The *thinnest freestanding* single-walled carbon nanotube is about 0.43 nm in diameter. Researchers suggested that it can be either (5,1) or (4,2) SWCNT, but the exact type of carbon nanotube remains questionable. (3,3), (4,3) and (5,1) carbon nanotubes (all about 0.4 nm in diameter) were unambiguously identified using aberration-corrected high-resolution transmission electron microscopy inside double-walled CNTs.

The *highest density* of CNTs was achieved in 2013, grown on a conductive titanium-coated copper surface that was coated with co-catalysts cobalt and molybdenum at lower than typical temperatures of 450 °C. The tubes averaged a height of 380 nm and a mass density of $1.6\,g\,cm^{-3}$. The material showed ohmic conductivity (lowest resistance $\sim22\,k\Omega$).

## Properties

## Strength

Carbon nanotubes are the strongest and stiffest materials yet discovered in terms of tensile strength and elastic modulus respectively. This strength results from the covalent $sp^2$ bonds formed between the individual carbon atoms. In 2000, a multi-walled carbon nanotube was tested to have a tensile strength of 63 gigapascals (9,100,000 psi). (For illustration, this translates into the ability to endure tension of a weight equivalent to 6,422 kilograms-force (62,980 N; 14,160 lbf) on a cable with cross-section of 1 square millimetre (0.0016 sq in).) Further studies, such as one conducted in 2008, revealed that individual CNT shells have strengths of up to ~100 gigapascals (15,000,000 psi), which is in agreement with quantum/atomistic models. Since carbon nanotubes have a low density for a solid of 1.3 to 1.4 g/cm$^3$, its specific strength of up to 48,000 kN·m·kg$^{-1}$ is the best of known materials, compared to high-carbon steel's 154 kN·m·kg$^{-1}$.

Under excessive tensile strain, the tubes will undergo plastic deformation, which means the deformation is permanent. This deformation begins at strains of approximately 5% and can increase the maximum strain the tubes undergo before fracture by releasing strain energy.

Although the strength of individual CNT shells is extremely high, weak shear interactions between adjacent shells and tubes lead to significant reduction in the effective strength of multi-walled carbon nanotubes and carbon nanotube bundles down to only a few GPa. This limitation has been recently addressed by applying high-energy electron irradiation, which crosslinks inner shells and tubes, and effectively increases the strength of these materials to ~60 GPa for multi-walled carbon nanotubes and ~17 GPa for double-walled carbon nanotube bundles.

CNTs are not nearly as strong under compression. Because of their hollow structure and high aspect ratio, they tend to undergo buckling when placed under compressive, torsional, or bending stress.

Comparison of mechanical properties

| Material | Young's modulus (TPa) | Tensile strength (GPa) | Elongation at break (%) |
|---|---|---|---|
| SWNT[E] | ~1 (from 1 to 5) | 13–53 | 16 |
| Armchair SWNT[T] | 0.94 | 126.2 | 23.1 |
| Zigzag SWNT[T] | 0.94 | 94.5 | 15.6–17.5 |
| Chiral SWNT | 0.92 | | |
| MWNT[E] | 0.2–0.8–0.95 | 11–63–150 | |
| Stainless steel[E] | 0.186–0.214 | 0.38–1.55 | 15–50 |
| Kevlar–29&149[E] | 0.06–0.18 | 3.6–3.8 | ~2 |

[E]Experimental observation; [T]Theoretical prediction

The above discussion referred to axial properties of the nanotube, whereas simple geometrical considerations suggest that carbon nanotubes should be much softer in the radial direction than along the tube axis. Indeed, TEM observation of radial elasticity suggested that even the van der Waals forces can deform two adjacent nanotubes. Nanoindentation experiments, performed by several groups on multiwalled carbon nanotubes and tapping/contact mode atomic force microscope measurements performed on single-walled carbon nanotubes, indicated a Young's modulus of the order of several GPa, confirming that CNTs are indeed rather soft in the radial direction.

## Hardness

Standard single-walled carbon nanotubes can withstand a pressure up to 25 GPa without [plastic/permanent] deformation. They then undergo a transformation to super-hard phase nanotubes. Maximum pressures measured using current experimental techniques are around 55 GPa. However, these new superhard phase nanotubes collapse at an even higher, albeit unknown, pressure.

The bulk modulus of superhard phase nanotubes is 462 to 546 GPa, even higher than that of diamond (420 GPa for single diamond crystal).

## Wettability

The surface wettability of CNT is of importance for its applications in various settings. Although the intrinsic contact angle of graphite is around 90°, the contact angles of most as-synthesized CNT arrays are over 160°, exhibiting a superhydrophobic property. By applying a low voltage as low as 1.3V, the extreme water repellant surface can be switched into superhydrophilic.

## Kinetic Properties

Multi-walled nanotubes are multiple concentric nanotubes precisely nested within one another. These exhibit a striking telescoping property whereby an inner nanotube core may slide, almost without friction, within its outer nanotube shell, thus creating an atomically perfect linear or rotational bearing. This is one of the first true examples of molecular nanotechnology, the precise positioning of atoms to create useful machines. Already, this property has been utilized to create the world's smallest rotational motor. Future applications such as a gigahertz mechanical oscillator are also envisioned.

## Electrical Properties

Unlike graphene, which is a two-dimensional semimetal, carbon nanotubes are either metallic or semiconducting along the tubular axis. For a given $(n,m)$ nanotube, if $n = m$, the nanotube is metallic; if $n - m$ is a multiple of 3, then the nanotube is semiconducting with a very small band gap, otherwise the nanotube is a moderate semiconductor. Thus all armchair ($n = m$) nanotubes are metallic, and nanotubes (6,4), (9,1), etc. are semiconducting. Carbon nanotubes are not semimetallic because the degenerate point (that point where the $\pi$ [bonding] band meets the $\pi^*$ [anti-bonding] band, at which the energy goes to zero) is slightly shifted away from the $K$ point in the Brillouin zone due to the curvature of the tube surface, casing hybridization between the $\sigma^*$ and $\pi^*$ anti-bonding bands, modifying the band dispersion.

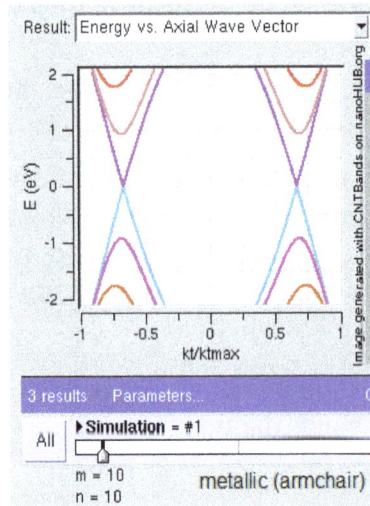

Band structures computed using tight binding approximation for (6,0) CNT (zigzag, metallic), (10,2) CNT (semiconducting) and (10,10) CNT (armchair, metallic).

The rule regarding metallic versus semiconductor behavior has exceptions, because curvature effects in small diameter tubes can strongly influence electrical properties. Thus, a (5,0) SWCNT that should be semiconducting in fact is metallic according to the calculations. Likewise, zigzag and chiral SWCNTs with small diameters that should be metallic have a finite gap (armchair nanotubes remain metallic). In theory, metallic nanotubes can carry an electric current density of $4 \times 10^9$ A/cm², which is more than 1,000 times greater than those of metals such as copper, where for copper intercon- nects current densities are limited by electromigration. Carbon nanotubes are thus be- ing explored as conductivity enhancing components in composite materials and many groups are attempting to commercialize highly conducting electrical wire assembled from individual carbon nanotubes. There are significant challenges to be overcome, however, such as the much more resistive nanotube-to-nanotube junctions and impu- rities, all of which lower the electrical conductivity of the macroscopic nanotube wires by orders of magnitude, as compared to the conductivity of the individual nanotubes.

Because of its nanoscale cross-section, electrons propagate only along the tube's axis. As a result, carbon nanotubes are frequently referred to as one-dimensional conduc- tors. The maximum electrical conductance of a single-walled carbon nanotube is $2G_0$, where $G_0 = 2e^2/h$ is the conductance of a single ballistic quantum channel.

Due to the role of the $\pi$-electron system in determining the electronic properties of graphene, doping in carbon nanotubes differs from that of bulk crystalline semicon- ductors from the same group of the periodic table (e.g. silicon). Graphitic substitution of carbon atoms in the nanotube wall by boron or nitrogen dopants leads to p-type and n-type behavior, respectively, as would be expected in silicon. However, some non-substitutional (intercalated or adsorbed) dopants introduced into a carbon nanotube, such as alkali metals as well as electron-rich metallocenes, result in n-type conduction because they donate electrons to the $\pi$-electron system of the nanotube. By contrast, $\pi$-electron acceptors such as $FeCl_3$ or electron-deficient metallocenes function as p-type dopants since they draw $\pi$-electrons away from the top of the valence band.

Intrinsic superconductivity has been reported, although other experiments found no evidence of this, leaving the claim a subject of debate.

## Optical Properties

## Thermal Properties

All nanotubes are expected to be very good thermal conductors along the tube, exhibiting a property known as "ballistic conduction", but good insulators lateral to the tube axis. Measurements show that a SWNT has a room-temperature thermal conductivity along its axis of about 3500 $W \cdot m^{-1} \cdot K^{-1}$; compare this to copper, a metal well known for its good thermal conductivity, which transmits 385 $W \cdot m^{-1} \cdot K^{-1}$. A SWNT has a room-temperature thermal conductivity across its axis (in the radial direction) of about 1.52 $W \cdot m^{-1} \cdot K^{-1}$, which is about as thermally conductive as soil. The temperature stability of carbon nanotubes is estimated to be up to 2800 °C in vacuum and about 750 °C in air.

## Defects

As with any material, the existence of a crystallographic defect affects the material properties. Defects can occur in the form of atomic vacancies. High levels of such defects can lower the tensile strength by up to 85%. An important example is the Stone Wales defect, which creates a pentagon and heptagon pair by rearrangement of the bonds. Because of the very small structure of CNTs, the tensile strength of the tube is dependent on its weakest segment in a similar manner to a chain, where the strength of the weakest link becomes the maximum strength of the chain.

Crystallographic defects also affect the tube's electrical properties. A common result is lowered conductivity through the defective region of the tube. A defect in armchair-type tubes (which can conduct electricity) can cause the surrounding region to become semi-conducting, and single monatomic vacancies induce magnetic properties.

Crystallographic defects strongly affect the tube's thermal properties. Such defects lead to phonon scattering, which in turn increases the relaxation rate of the phonons. This reduces the mean free path and reduces the thermal conductivity of nanotube structures. Phonon transport simulations indicate that substitutional defects such as nitrogen or boron will primarily lead to scattering of high-frequency optical phonons. However, larger-scale defects such as Stone Wales defects cause phonon scattering over a wide range of frequencies, leading to a greater reduction in thermal conductivity.

## Safety and Health

The toxicity of carbon nanotubes has been an important question in nanotechnology. As of 2007, such research had just begun. The data is still fragmentary and subject to criticism. Preliminary results highlight the difficulties in evaluating the toxicity of this heterogeneous material. Parameters such as structure, size distribution, surface area, surface chemistry, surface charge, and agglomeration state as well as purity of the samples, have considerable impact on the reactivity of carbon nanotubes. However, available data clearly show that, under some conditions, nanotubes can cross membrane barriers, which suggests that, if raw materials reach the organs, they can induce harmful effects such as inflammatory and fibrotic reactions.

## Synthesis

Techniques have been developed to produce nanotubes in sizable quantities, including arc discharge, laser ablation, high-pressure carbon monoxide disproportionation, and chemical vapor deposition (CVD). Most of these processes take place in a vacuum or with process gases. CVD growth of CNTs can occur in vacuum or at atmospheric pressure. Large quantities of nanotubes can be synthesized by these methods; advances in catalysis and continuous growth are making CNTs more commercially viable.

## Chemical Modification

Carbon nanotubes can be functionalized to attain desired properties that can be used in a wide variety of applications. The two main methods of carbon nanotube functionalization are covalent and non-covalent modifications. Because of their hydrophobic nature, carbon nanotubes tend to agglomerate hindering their dispersion is solvents or viscous polymer melts. The resulting nanotube bundles or aggregates reduce the mechanical performance of the final composite. The surface of the carbon nanotubes can be modified to reduce the hydrophobicity and improve interfacial adhesion to a bulk polymer through chemical attachment.

## Applications

### Current

Current use and application of nanotubes has mostly been limited to the use of bulk nanotubes, which is a mass of rather unorganized fragments of nanotubes. Bulk nanotube materials may never achieve a tensile strength similar to that of individual tubes, but such composites may, nevertheless, yield strengths sufficient for many applications. Bulk carbon nanotubes have already been used as composite fibers in polymers to improve the mechanical, thermal and electrical properties of the bulk product.

- Easton-Bell Sports, Inc. have been in partnership with Zyvex Performance Materials, using CNT technology in a number of their bicycle components—including flat and riser handlebars, cranks, forks, seatposts, stems and aero bars.

- Zyvex Technologies has also built a 54' maritime vessel, the Piranha Unmanned Surface Vessel, as a technology demonstrator for what is possible using CNT technology. CNTs help improve the structural performance of the vessel, resulting in a lightweight 8,000 lb boat that can carry a payload of 15,000 lb over a range of 2,500 miles.

- Amroy Europe Oy manufactures Hybtonite carbon nanoepoxy resins where carbon nanotubes have been chemically activated to bond to epoxy, resulting in a composite material that is 20% to 30% stronger than other composite materials. It has been used for wind turbines, marine paints and variety of sports gear such as skis, ice hockey sticks, baseball bats, hunting arrows, and surfboards.

Other current applications include:

- tips for atomic force microscope probes

- in tissue engineering, carbon nanotubes can act as scaffolding for bone growth

There is also ongoing research in using carbon nanotubes as a scaffold for diverse microfabrication techniques.

## Potential

The strength and flexibility of carbon nanotubes makes them of potential use in controlling other nanoscale structures, which suggests they will have an important role in nanotechnology engineering. The highest tensile strength of an individual multi-walled carbon nanotube has been tested to be 63 GPa. Carbon nanotubes were found in Damascus steel from the 17th century, possibly helping to account for the legendary strength of the swords made of it. Recently, several studies have highlighted the prospect of using carbon nanotubes as building blocks to fabricate three-dimensional macroscopic (>1mm in all three dimensions) all-carbon devices. Lalwani et al. have reported a novel radical initiated thermal crosslinking method to fabricated macroscopic, free-standing, porous, all-carbon scaffolds using single- and multi-walled carbon nanotubes as building blocks. These scaffolds possess macro-, micro-, and nano- structured pores and the porosity can be tailored for specific applications. These 3D all-carbon scaffolds/architectures maybe used for the fabrication of the next generation of energy storage, supercapacitors, field emission transistors, high-performance catalysis, photovoltaics, and biomedical devices and implants.

## Discovery

The true identity of the discoverers of carbon nanotubes is a subject of some controversy. For years, scientists assumed that Sumio Iijima of NEC had discovered carbon
nanotubes in 1991. He published a paper describing his discovery which initiated
a flurry of excitement and could be credited by inspiring the many scientists now
studying applications of carbon nanotubes. Though Iijima has been given much of
the credit for discovering carbon nanotubes, it turns out that the timeline of carbon
nanotubes goes back much further than 1991. In 1952 L. V. Radushkevich and V. M.
Lukyanovich published clear images of 50 nanometer diameter tubes made of carbon
in the Soviet *Journal of Physical Chemistry*. This discovery was largely unnoticed, as
the article was published in Russian, and Western scientists' access to Soviet press
was limited during the Cold War. Before they came to be known as carbon nanotubes,
in 1976, Morinobu Endo of CNRS observed hollow tubes of rolled up graphite sheets
synthesised by a chemical vapour-growth technique. The first specimens observed
would later come to be known as single-walled carbon nanotubes (SWNTs). The
three scientists have been the first ones to show images of a nanotube with a solitary
graphene wall.

Endo, in his early review of vapor-phase-grown carbon fibers (VPCF), also reminded us
that he had observed a hollow tube, linearly extended with parallel carbon layer faces
near the fiber core. This appears to be the observation of multi-walled carbon nanotubes at the center of the fiber. The mass-produced MWCNTs today are strongly related
to the VPGCF developed by Endo. In fact, they call it the "Endo-process", out of respect
for his early work and patents.

In 1979, John Abrahamson presented evidence of carbon nanotubes at the 14th Bi-
ennial Conference of Carbon at Pennsylvania State University. The conference paper
described carbon nanotubes as carbon fibers that were produced on carbon anodes
during arc discharge. A characterization of these fibers was given as well as hypotheses
for their growth in a nitrogen atmosphere at low pressures.

In 1981, a group of Soviet scientists published the results of chemical and structural
characterization of carbon nanoparticles produced by a thermocatalytical disproportionation of carbon monoxide. Using TEM images and XRD patterns, the authors suggested that their "carbon multi-layer tubular crystals" were formed by rolling graphene
layers into cylinders. They speculated that by rolling graphene layers into a cylinder,
many different arrangements of graphene hexagonal nets are possible. They suggested
two possibilities of such arrangements: circular arrangement (armchair nanotube) and
a spiral, helical arrangement (chiral tube).

In 1987, Howard G. Tennent of Hyperion Catalysis was issued a U.S. patent for the
production of "cylindrical discrete carbon fibrils" with a "constant diameter between
about 3.5 and about 70 nanometers, length $10^2$ times the diameter, and an outer region

of multiple essentially continuous layers of ordered carbon atoms and a distinct inner core."

Iijima's discovery of multi-walled carbon nanotubes in the insoluble material of arc-burned graphite rods in 1991 and Mintmire, Dunlap, and White's independent prediction that if single-walled carbon nanotubes could be made, then they would exhibit remarkable conducting properties helped create the initial buzz that is now associated with carbon nanotubes. Nanotube research accelerated greatly following the independent discoveries by Bethune at IBM and Iijima at NEC of *single-walled* carbon nanotubes and methods to specifically produce them by adding transition-metal catalysts to the carbon in an arc discharge. The arc discharge technique was well-known to produce the famed Buckminster fullerene on a preparative scale, and these results appeared to extend the run of accidental discoveries relating to fullerenes. The discovery of nanotubes remains a contentious issue. Many believe that Iijima's report in 1991 is of particular importance because it brought carbon nanotubes into the awareness of the scientific community as a whole.

# Fullerene

A fullerene is a molecule of carbon in the form of a hollow sphere, ellipsoid, tube, and many other shapes. Spherical fullerenes, also referred to as Buckminsterfullerenes (buckyballs), resemble the balls used in football (soccer). Cylindrical fullerenes are also called carbon nanotubes (buckytubes). Fullerenes are similar in structure to graphite, which is composed of stacked graphene sheets of linked hexagonal rings; they may also contain pentagonal (or sometimes heptagonal) rings.

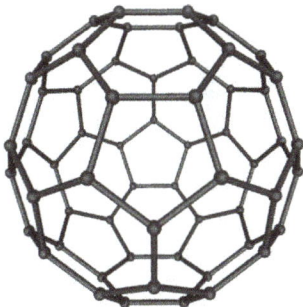

Buckminsterfullerene $C_{60}$ (left) and carbon nanotubes (right) are two examples of structures in the fullerene family.

The first fullerene molecule to be discovered, and the family's namesake, buckminsterfullerene ($C_{60}$), was manufactured in 1985 by Richard Smalley, Robert Curl, James Heath, Sean O'Brien, and Harold Kroto at Rice University. The name was an homage to Buckminster Fuller, whose geodesic domes it resembles. The structure was also identified some five years earlier by Sumio Iijima, from an electron microscope image, where it formed the core of a "bucky onion". Fullerenes have since been found to occur in nature. More recently, fullerenes have been detected in outer space. According to astronomer Letizia Stanghellini, "It's possible that buckyballs from outer space provided seeds for life on Earth."

The discovery of fullerenes greatly expanded the number of known carbon allotropes, which until recently were limited to graphite, graphene, diamond, and amorphous carbon such as soot and charcoal. Buckyballs and buckytubes have been the subject of intense research, both for their unique chemistry and for their technological applications, especially in materials science, electronics, and nanotechnology.

## History

The icosahedral $C_{60}H_{60}$ cage was mentioned in 1965 as a possible topological structure. Eiji Osawa of Toyohashi University of Technology predicted the existence of $C_{60}$ in 1970. He noticed that the structure of a corannulene molecule was a subset of an Association football shape, and he hypothesised that a full ball shape could also exist. Japanese scientific journals reported his idea, but neither it nor any translations of it reached Europe or the Americas.

The icosahedral fullerene $C_{540}$, another member of the family of fullerenes.

Also in 1970, R. W. Henson (then of the Atomic Energy Research Establishment) proposed the structure and made a model of $C_{60}$. Unfortunately, the evidence for this new form of carbon was very weak and was not accepted, even by his colleagues. The results were never published but were acknowledged in *Carbon* in 1999.

In 1973 independently from Henson, a group of scientists from the USSR, directed by Prof. Bochvar, made a quantum-chemical analysis of the stability of $C_{60}$ and calculated its electronic structure. As in the previous cases, the scientific community did not accept the theoretical prediction. The paper was published in 1973 in *Proceedings of the USSR Academy of Sciences* (in Russian).

In mass spectrometry discrete peaks appeared corresponding to molecules with the exact mass of sixty or seventy or more carbon atoms. In 1985 Harold Kroto of the University of

Sussex, James R. Heath, Sean O'Brien, Robert Curl and Richard Smalley from Rice University, discovered $C_{60}$, and shortly thereafter came to discover the fullerenes. Kroto, Curl, and Smalley were awarded the 1996 Nobel Prize in Chemistry for their roles in the discovery of this class of molecules. $C_{60}$ and other fullerenes were later noticed occurring outside the laboratory (for example, in normal candle-soot). By 1990 it was relatively easy to produce gram-sized samples of fullerene powder using the techniques of Donald Huffman, Wolfgang Krätschmer, Lowell D. Lamb, and Konstantinos Fostiropoulos. Fullerene purification remains a challenge to chemists and to a large extent determines fullerene prices. So-called endohedral fullerenes have ions or small molecules incorporated inside the cage atoms. Fullerene is an unusual reactant in many organic reactions such as the Bingel reaction discovered in 1993. Carbon nanotubes were first discovered and synthesized in 1991.

Minute quantities of the fullerenes, in the form of $C_{60}$, $C_{70}$, $C_{76}$, $C_{82}$ and $C_{84}$ molecules, are produced in nature, hidden in soot and formed by lightning discharges in the atmosphere. In 1992, fullerenes were found in a family of minerals known as Shungites in Karelia, Russia.

In 2010, fullerenes ($C_{60}$) have been discovered in a cloud of cosmic dust surrounding a distant star 6500 light years away. Using NASA's Spitzer infrared telescope the scientists spotted the molecules' unmistakable infrared signature. Sir Harry Kroto, who shared the 1996 Nobel Prize in Chemistry for the discovery of buckyballs commented: "This most exciting breakthrough provides convincing evidence that the buckyball has, as I long suspected, existed since time immemorial in the dark recesses of our galaxy."

## Naming

The discoverers of the Buckminsterfullerene ($C_{60}$) allotrope of carbon named it after Richard Buckminster Fuller, a noted architectural modeler who popularized the geodesic dome. Since buckminsterfullerenes have a similar shape to those of such domes, they thought the name appropriate. As the discovery of the fullerene family came *after* buckminsterfullerene, the shortened name 'fullerene' is used to refer to the family of fullerenes. The suffix "-ene" indicates that each C atom is covalently bonded to three others (instead of the maximum of four), a situation that classically would correspond to the existence of bonds involving two pairs of electrons ("double bonds").

## Types of Fullerene

Since the discovery of fullerenes in 1985, structural variations on fullerenes have evolved well beyond the individual clusters themselves. Examples include:

- Buckyball clusters: smallest member is C 20 (unsaturated version of dodecahedrane) and the most common is C 60;

- Nanotubes: hollow tubes of very small dimensions, having single or multiple walls; potential applications in electronics industry;

- Megatubes: larger in diameter than nanotubes and prepared with walls of different thickness; potentially used for the transport of a variety of molecules of different sizes;

- polymers: chain, two-dimensional and three-dimensional polymers are formed under high-pressure high-temperature conditions; single-strand polymers are formed using the Atom Transfer Radical Addition Polymerization (ATRAP) route;

- nano"onions": spherical particles based on multiple carbon layers surrounding a buckyball core; proposed for lubricants;

- linked "ball-and-chain" dimers: two buckyballs linked by a carbon chain;

- fullerene rings.

## Buckyballs

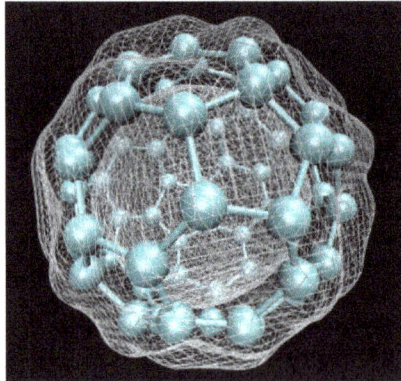

$C_{60}$ with isosurface of ground state electron density as calculated with DFT

## Buckminsterfullerene

Buckminsterfullerene is the smallest fullerene molecule containing pentagonal and hexagonal rings in which no two pentagons share an edge (which can be destabilizing, as in pentalene). It is also most common in terms of natural occurrence, as it can often be found in soot.

The structure of $C_{60}$ is a truncated icosahedron, which resembles an association football ball of the type made of twenty hexagons and twelve pentagons, with a carbon atom at the vertices of each polygon and a bond along each polygon edge.

The van der Waals diameter of a $C_{60}$ molecule is about 1.1 nanometers (nm). The nucleus to nucleus diameter of a $C_{60}$ molecule is about 0.71 nm.

The $C_{60}$ molecule has two bond lengths. The 6:6 ring bonds (between two hexagons) can be considered "double bonds" and are shorter than the 6:5 bonds (between a hexagon and a pentagon). Its average bond length is 1.4 angstroms.

Silicon buckyballs have been created around metal ions.

## Boron Buckyball

A type of buckyball which uses boron atoms, instead of the usual carbon, was predicted and described in 2007. The $B_{80}$ structure, with each atom forming 5 or 6 bonds, is predicted to be more stable than the $C_{60}$ buckyball. One reason for this given by the researchers is that the B-80 is actually more like the original geodesic dome structure popularized by Buckminster Fuller, which uses triangles rather than hexagons. However, this work has been subject to much criticism by quantum chemists as it was concluded that the predicted $I_h$ symmetric structure was vibrationally unstable and the resulting cage undergoes a spontaneous symmetry break, yielding a puckered cage with rare $T_h$ symmetry (symmetry of a volleyball). The number of six-member rings in this molecule is 20 and number of five-member rings is 12. There is an additional atom in the center of each six-member ring, bonded to each atom surrounding it. By employing a systematic global search algorithm, later it was found that the previously proposed B80 fullerene is not global minimum for 80 atom boron clusters and hence can not be found in nature. In the same paper by Sandip De et al., it was concluded that boron's energy landscape is significantly different from other fullerenes already found in nature hence pure boron fullerenes are unlikely to exist in nature.

## Other Buckyballs

Another fairly common fullerene is $C_{70}$, but fullerenes with 72, 76, 84 and even up to 100 carbon atoms are commonly obtained.

In mathematical terms, the structure of a fullerene is a trivalent convex polyhedron with pentagonal and hexagonal faces. In graph theory, the term fullerene refers to any 3-regular, planar graph with all faces of size 5 or 6 (including the external face). It follows from Euler's polyhedron formula, $V - E + F = 2$ (where $V$, $E$, $F$ are the numbers of vertices, edges, and faces), that there are exactly 12 pentagons in a fullerene and $V/2 - 10$ hexagons.

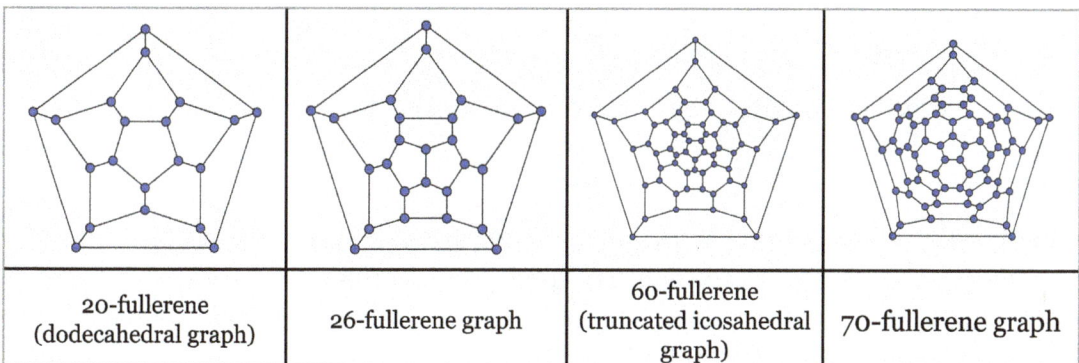

| 20-fullerene (dodecahedral graph) | 26-fullerene graph | 60-fullerene (truncated icosahedral graph) | 70-fullerene graph |
| --- | --- | --- | --- |

The smallest fullerene is the dodecahedral $C_{20}$. There are no fullerenes with 22 vertices. The number of fullerenes $C_{2n}$ grows with increasing $n = 12, 13, 14, ...$, roughly in proportion to $n^9$ (sequence A007894 in the OEIS). For instance, there are 1812 non-isomorphic fullerenes $C_{60}$. Note that only one form of $C_{60}$, the buckminsterfullerene alias truncated icosahedron, has no pair of adjacent pentagons (the smallest such fullerene). To further illustrate the growth, there are 214,127,713 non-isomorphic fullerenes $C_{200}$, 15,655,672 of which have no adjacent pentagons. Optimized structures of many fullerene isomers are published and listed on the web.

Heterofullerenes have heteroatoms substituting carbons in cage or tube-shaped structures. They were discovered in 1993 and greatly expand the overall fullerene class of compounds. Notable examples include boron, nitrogen (azafullerene), oxygen, and phosphorus derivatives.

Trimetasphere carbon nanomaterials were discovered by researchers at Virginia Tech and licensed exclusively to Luna Innovations. This class of novel molecules comprises 80 carbon atoms (C 80) forming a sphere which encloses a complex of three metal atoms and one nitrogen atom. These fullerenes encapsulate metals which puts them in the subset referred to as metallofullerenes. Trimetaspheres have the potential for use in diagnostics (as safe imaging agents), therapeutics and in organic solar cells.

## Carbon Nanotubes

Nanotubes are cylindrical fullerenes. These tubes of carbon are usually only a few nanometres wide, but they can range from less than a micrometer to several millimeters in length. They often have closed ends, but can be open-ended as well. There are also cases in which the tube reduces in diameter before closing off. Their unique molecular structure results in extraordinary macroscopic properties, including high tensile strength, high electrical conductivity, high ductility, high heat conductivity, and relative chemical inactivity (as it is cylindrical and "planar" — that is, it has no "exposed" atoms that can be easily displaced). One proposed use of carbon nanotubes is in paper batteries, developed in 2007 by researchers at Rensselaer Polytechnic Institute. Another highly speculative proposed use in the field of space technologies is to produce high-tensile carbon cables required by a space elevator.

## Carbon Nanobuds

Nanobuds have been obtained by adding buckminsterfullerenes to carbon nanotubes.

## Fullerite

"Ultrahard fullerite" is a coined term frequently used to describe material produced by high-pressure high-temperature (HPHT) processing of fullerite. Such treatment converts fullerite into a nanocrystalline form of diamond which has been reported to exhibit remarkable mechanical properties.

The $C_{60}$ fullerene in crystalline form
Fullerites are the solid-state manifestation of fullerenes and related compounds and materials.

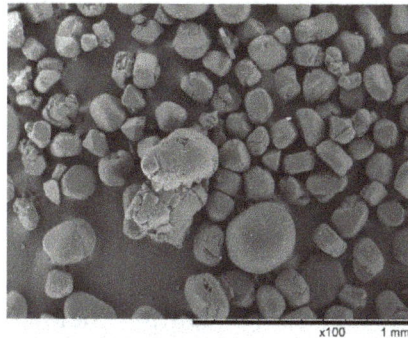

Fullerite (scanning electron microscope image)

## Inorganic Fullerenes

Materials with fullerene-like molecular structures but lacking carbon include $MoS_2$, $WS_2$, $TiS_2$ and $NbS_2$. Prof. J. M. Martin from Ecole Centrale de Lyon in France tested the new material under isostatic pressure and found it to be stable up to at least 350 tons/cm².

## Properties

For the past decade, the chemical and physical properties of fullerenes have been a hot topic in the field of research and development, and are likely to continue to be for a long time. *Popular Science* has discussed possible uses of fullerenes (graphene) in armor. In April 2003, fullerenes were under study for potential medicinal use: binding specific antibiotics to the structure to target resistant bacteria and even target certain cancer cells such as melanoma. The October 2005 issue of *Chemistry & Biology* contains an article describing the use of fullerenes as light-activated antimicrobial agents.

In the field of nanotechnology, heat resistance and superconductivity are some of the more heavily studied properties.

A common method used to produce fullerenes is to send a large current between two nearby graphite electrodes in an inert atmosphere. The resulting carbon plasma arc between the electrodes cools into sooty residue from which many fullerenes can be isolated.

There are many calculations that have been done using ab-initio quantum methods applied to fullerenes. By DFT and TD-DFT methods one can obtain IR, Raman and UV spectra. Results of such calculations can be compared with experimental results.

## Aromaticity

Researchers have been able to increase the reactivity of fullerenes by attaching active groups to their surfaces. Buckminsterfullerene does not exhibit "superaromaticity": that is, the electrons in the hexagonal rings do not delocalize over the whole molecule.

A spherical fullerene of $n$ carbon atoms has $n$ pi-bonding electrons, free to delocalize. These should try to delocalize over the whole molecule. The quantum mechanics of such an arrangement should be like one shell only of the well-known quantum mechanical structure of a single atom, with a stable filled shell for $n = 2, 8, 18, 32, 50, 72, 98, 128$, etc.; i.e. twice a perfect square number; but this series does not include 60. This $2(N + 1)^2$ rule (with $N$ integer) for spherical aromaticity is the three-dimensional analogue of Hückel's rule. The 10+ cation would satisfy this rule, and should be aromatic. This has been shown to be the case using quantum chemical modelling, which showed the existence of strong diamagnetic sphere currents in the cation.

As a result, $C_{60}$ in water tends to pick up two more electrons and become an anion. The $nC_{60}$ described below may be the result of $C_{60}$ trying to form a loose metallic bond.

## Chemistry

Fullerenes are stable, but not totally unreactive. The $sp^2$-hybridized carbon atoms, which are at their energy minimum in planar graphite, must be bent to form the closed sphere or tube, which produces angle strain. The characteristic reaction of fullerenes is electrophilic addition at 6,6-double bonds, which reduces angle strain by changing $sp^2$-hybridized carbons into $sp^3$-hybridized ones. The change in hybridized orbitals causes the bond angles to decrease from about 120° in the $sp^2$ orbitals to about 109.5° in the $sp^3$ orbitals. This decrease in bond angles allows for the bonds to bend less when closing the sphere or tube, and thus, the molecule becomes more stable.

Other atoms can be trapped inside fullerenes to form inclusion compounds known as endohedral fullerenes. An unusual example is the egg-shaped fullerene $Tb_3N@C_{84}$, which violates the isolated pentagon rule. Recent evidence for a meteor impact at the end of the Permian period was found by analyzing noble gases so preserved. Metallofullerene-based inoculates using the rhonditic steel process are beginning production as one of the first commercially viable uses of buckyballs.

# Solubility

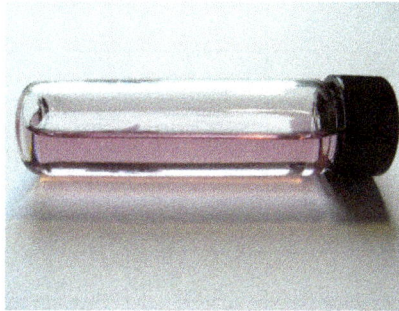

$C_{60}$ in solution

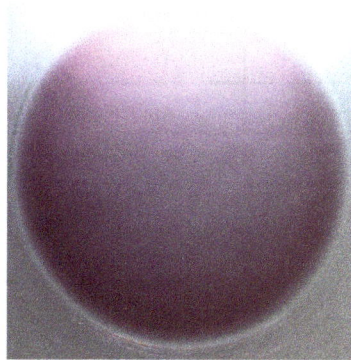

$C_{60}$ in extra virgin olive oil showing the characteristic purple color of pristine $C_{60}$ solutions.

Fullerenes are sparingly soluble in many solvents. Common solvents for the fullerenes include aromatics, such as toluene, and others like carbon disulfide. Solutions of pure buckminsterfullerene have a deep purple color. Solutions of $C_{70}$ are a reddish brown. The higher fullerenes $C_{76}$ to $C_{84}$ have a variety of colors. $C_{76}$ has two optical forms, while other higher fullerenes have several structural isomers. Fullerenes are the only known allotrope of carbon that can be dissolved in common solvents at room temperature.

| Solvent | $C_{60}$ mg/mL | $C_{70}$ mg/mL |
|---|---|---|
| 1-chloronaphthalene | 51 | ND |
| 1-methylnaphthalene | 33 | ND |
| 1,2-dichlorobenzene | 24 | 36.2 |
| 1,2,4-trimethylbenzene | 18 | ND |
| tetrahydronaphthalene | 16 | ND |
| carbon disulfide | 8 | 9.875 |
| 1,2,3-tribromopropane | 8 | ND |
| chlorobenzene | 7 | ND |
| p-xylene | 5 | 3.985 |

| Solvent | $C_{60}$ mg/mL | $C_{70}$ mg/mL |
|---|---|---|
| bromoform | 5 | ND |
| cumene | 4 | ND |
| toluene | 3 | 1.406 |
| benzene | 1.5 | 1.3 |
| carbon tetrachloride | 0.447 | 0.121 |
| chloroform | 0.25 | ND |
| n-hexane | 0.046 | 0.013 |
| cyclohexane | 0.035 | 0.08 |
| tetrahydrofuran | 0.006 | ND |
| acetonitrile | 0.004 | ND |
| methanol | $4.0 \times 10^{-5}$ | ND |
| water | $1.3 \times 10^{-11}$ | ND |
| pentane | 0.004 | 0.002 |
| heptane | ND | 0.047 |
| octane | 0.025 | 0.042 |
| isooctane | 0.026 | ND |
| decane | 0.070 | 0.053 |
| dodecane | 0.091 | 0.098 |
| tetradecane | 0.126 | ND |
| acetone | ND | 0.0019 |
| isopropanol | ND | 0.0021 |
| dioxane | 0.0041 | ND |
| mesitylene | 0.997 | 1.472 |
| dichloromethane | 0.254 | 0.080 |
| ND, not determined | | |

Some fullerene structures are not soluble because they have a small band gap between the ground and excited states. These include the small fullerenes $C_{28}$, $C_{36}$ and $C_{50}$. The $C_{72}$ structure is also in this class, but the endohedral version with a trapped lanthanide-group atom is soluble due to the interaction of the metal atom and the electronic states of the fullerene. Researchers had originally been puzzled by $C_{72}$ being absent in fullerene plasma-generated soot extract, but found in endohedral samples. Small band gap fullerenes are highly reactive and bind to other fullerenes or to soot particles.

Solvents that are able to dissolve buckminsterfullerene ($C_{60}$ and $C_{70}$) are listed at left in order from highest solubility. The solubility value given is the approximate saturated concentration.

Solubility of $C_{60}$ in some solvents shows unusual behaviour due to existence of solvate phases (analogues of crystallohydrates). For example, solubility of $C_{60}$ in benzene solution shows maximum at about 313 K. Crystallization from benzene solution at temperatures below maximum results in formation of triclinic solid solvate with four benzene molecules $C_{60} \cdot 4C_6H_6$ which is rather unstable in air. Out of solution, this structure decomposes into usual face-centered cubic (fcc) $C_{60}$ in few minutes' time. At temperatures above solubility maximum the solvate is not stable even when immersed in saturated solution and melts with formation of fcc $C_{60}$. Crystallization at temperatures above the solubility maximum results in formation of pure fcc $C_{60}$. Millimeter-sized crystals of $C_{60}$ and $C_{70}$ can be grown from solution both for solvates and for pure fullerenes.

## Quantum Mechanics

In 1999, researchers from the University of Vienna demonstrated that wave-particle duality applied to molecules such as fullerene.

## Superconductivity

## Chirality

Some fullerenes (e.g. $C_{76}$, $C_{78}$, $C_{80}$, and $C_{84}$) are inherently chiral because they are $D_2$-symmetric, and have been successfully resolved. Research efforts are ongoing to develop specific sensors for their enantiomers.

## Construction

Two theories have been proposed to describe the molecular mechanisms that make fullerenes. The older, "bottom-up" theory proposes that they are built atom-by-atom. The alternative "top-down" approach claims that fullerenes form when much larger structures break into constituent parts.

In 2013 researchers discovered that asymmetrical fullerenes formed from larger structures settle into stable fullerenes. The synthesized substance was a particular metallofullerene consisting of 84 carbon atoms with two additional carbon atoms and two yttrium atoms inside the cage. The process produced approximately 100 micrograms.

However, they found that the asymmetrical molecule could theoretically collapse to form nearly every known fullerene and metallofullerene. Minor perturbations involving the breaking of a few molecular bonds cause the cage to become highly symmetrical and stable. This insight supports the theory that fullerenes can be formed from graphene when the appropriate molecular bonds are severed.

## Production Technology

Fullerene production processes comprise the following five subprocesses: (i) synthesis of fullerenes or fullerene-containing soot; (ii) extraction; (iii) separation (purification) for each fullerene molecule, yielding pure fullerenes such as $C_{60}$; (iv) synthesis of derivatives (mostly using the techniques of organic synthesis); (v) other post-processing such as dispersion into a matrix. The two synthesis methods used in practice are the arc method, and the combustion method. The latter, discovered at the Massachusetts Institute of Technology, is preferred for large scale industrial production.

## Applications

Fullerenes have been extensively used for several biomedical applications including the design of high-performance MRI contrast agents, X-Ray imaging contrast agents, photodynamic therapy and drug and gene delivery, summarized in several comprehensive reviews.

## Tumor Research

While past cancer research has involved radiation therapy, photodynamic therapy is important to study because breakthroughs in treatments for tumor cells will give more options to patients with different conditions. More recent experiments using HeLa cells in cancer research involves the development of new photosensitizers with increased ability to be absorbed by cancer cells and still trigger cell death. It is also important that a new photosensitizer does not stay in the body for a long time to prevent unwanted cell damage.

Fullerenes can be made to be absorbed by HeLa cells. The $C_{60}$ derivatives can be delivered to the cells by using the functional groups L-phenylalanine, folic acid, and L-arginine among others. The purpose for functionalizing the fullerenes is to increase the solubility of the molecule by the cancer cells. Cancer cells take up these molecules at an increased rate because of an upregulation of transporters in the cancer cell, in this case amino acid transporters will bring in the L-arginine and L-phenylalanine functional groups of the fullerenes.

Once absorbed by the cells, the $C_{60}$ derivatives would react to light radiation by turning molecular oxygen into reactive oxygen which triggers apoptosis in the HeLa cells and other cancer cells that can absorb the fullerene molecule. This research shows that a reactive substance can target cancer cells and then be triggered by light radiation, minimizing damage to surrounding tissues while undergoing treatment.

When absorbed by cancer cells and exposed to light radiation, the reaction that creates reactive oxygen damages the DNA, proteins, and lipids that make up the cancer cell. This cellular damage forces the cancerous cell to go through apoptosis, which can lead to the reduction in size of a tumor. Once the light radiation treatment is finished the fullerene will reabsorb the free radicals to prevent damage of other tissues. Since this treatment focuses on cancer cells, it is a good option for patients whose cancer cells are within reach of light radiation. As this research continues into the future, it will be able to penetrate deeper into the body and be absorbed by cancer cells more effectively.

## Safety and Toxicity

A comprehensive and recent review on fullerene toxicity is given by Lalwani et al. These authors review the works on fullerene toxicity beginning in the early 1990s to present, and conclude that very little evidence gathered since the discovery of fullerenes indicate that $C_{60}$ is toxic. The toxicity of these carbon nanoparticles is not only dose and time-dependent, but also depends on a number of other factors such as: type (e.g., $C_{60}$, $C_{70}$, $M@C_{60}$, $M@C_{82}$, functional groups used to water solubilize these nanoparticles (e.g., OH, COOH), and method of administration (e.g., intravenous, intraperitoneal). The authors therefore recommend that pharmacology of every new fullerene- or metallofullerene-based complex must be assessed individually as a different compound.

## Popular Culture

Examples of fullerenes in popular culture are numerous. Fullerenes appeared in fiction well before scientists took serious interest in them. In a humorously speculative 1966 column for *New Scientist*, David Jones suggested that it may be possible to create giant hollow carbon molecules by distorting a plane hexagonal net by the addition of impurity atoms.

On 4 September 2010, Google used an interactively rotatable fullerene $C_{60}$ as the second 'o' in their logo to celebrate the 25th anniversary of the discovery of the fullerenes.

## Nanosensor

Nanosensors are any biological, chemical, or surgical sensory points used to convey information about nanoparticles to the macroscopic world. Their use mainly include various medicinal purposes and as gateways to building other nanoproducts, such as computer chips that work at the nanoscale and nanorobots. There are several ways being proposed today to make nanosensors; these include top-down lithography, bottom-up assembly, and molecular self-assembly.

## Predicted Applications

Medicinal uses of nanosensors mainly revolve around the potential of nanosensors to accurately identify particular cells or places in the body in need. By measuring changes in volume, concentration, displacement and velocity, gravitational, electrical, and magnetic forces, pressure, or temperature of cells in a body, nanosensors may be able to distinguish between and recognize certain cells, most notably those of cancer, at the molecular level in order to deliver medicine or monitor development to specific places in the body. For example, light-responsive nanosensors can be designed to determine local protease activity in a body to detect immune response or cancer. In addition, they may be able to detect macroscopic variations from outside the body and communicate these changes to other nanoproducts working within the body.

One example of nanosensors involves using the fluorescence properties of cadmium selenide quantum dots as sensors to uncover tumors within the body. By injecting a body with these quantum dots, a doctor could see where a tumor or cancer cell was by finding the injected quantum dots, an easy process because of their fluorescence. Developed nanosensor quantum dots would be specifically constructed to find only the particular cell for which the body was at risk. A downside to the cadmium selenide dots, however, is that they are highly toxic to the body. As a result, researchers are working on developing alternate dots made out of a different, less toxic material while still re-taining some of the fluorescence properties. In particular, they have been investigating the particular benefits of zinc sulfide quantum dots which, though they are not quite as fluorescent as cadmium selenide, can be augmented with other metals including manganese and various lanthanide elements. In addition, these newer quantum dots become more fluorescent when they bond to their target cells. (Quantum) Potential predicted functions may also include sensors used to detect specific DNA in order to recognize explicit genetic defects, especially for individuals at high-risk and implanted sensors that can automatically detect glucose levels for diabetic subjects more simply than current detectors. DNA can also serve as sacrificial layer for manufacturing CMOS IC, integrating a nanodevice with sensing capabilities. Therefore, using proteomic pat-terns and new hybrid materials, nanobiosensors can also be used to enable components configured into a hybrid semiconductor substrate as part of the circuit assembly. The development and miniaturization of nanobiosensors should provide interesting new opportunities.

Other projected products most commonly involve using nanosensors to build smaller integrated circuits, as well as incorporating them into various other commodities made using other forms of nanotechnology for use in a variety of situations including trans-portation, communication, improvements in structural integrity, and robotics. Nano-sensors may also eventually be valuable as more accurate monitors of material states for use in systems where size and weight are constrained, such as in satellites and other aeronautic machines.

## Existing Examples

Currently, the most common mass-produced functioning nanosensors exist in the bio-logical world as natural receptors of outside stimulation. For instance, sense of smell, especially in animals in which it is particularly strong, such as dogs, functions using receptors that sense nanosized molecules. Certain plants, too, use nanosensors to detect sunlight; various fish use nanosensors to detect minuscule vibrations in the surrounding water; and many insects detect sex pheromones using nanosensors.

One of the first working examples of a synthetic nanosensor was built by researchers at the Georgia Institute of Technology in 1999. It involved attaching a single particle onto the end of a carbon nanotube and measuring the vibrational frequency of the nanotube both with and without the particle. The discrepancy between the two frequencies allowed the researchers to measure the mass of the attached particle.

Chemical sensors, too, have been built using nanotubes to detect various properties of gaseous molecules. Carbon nanotubes have been used to sense ionization of gaseous molecules while nanotubes made out of titanium have been employed to detect atmospheric concentrations of hydrogen at the molecular level. Many of these involve a system by which nanosensors are built to have a specific pocket for another molecule. When that particular molecule, and only that specific molecule, fits into the nanosensor, and light is shone upon the nanosensor, it will reflect different wavelengths of light and, thus, be a different color. In a similar fashion, Flood et al. have shown that supramolecular host-guest chemistry offers quantitative sensing using Raman scattered light as well as SERS.

Photonic devices can also be used as nanosensors to quantify concentrations of clinically relevant samples. A principle of operation of these sensors is based on the chemical modulation of a hydrogel film volume that incorporates a Bragg grating. As the hydrogel swells or shrinks upon chemical stimulation, the Bragg grating changes color and diffracts light at different wavelengths. The diffracted light can be correlated with the concentration of a target analyte.

## Production Methods

There are currently several hypothesized ways to produce nanosensors. Top-down lithography is the manner in which most integrated circuits are now made. It involves starting out with a larger block of some material and carving out the desired form. These carved out devices, notably put to use in specific microelectromechanical systems used as microsensors, generally only reach the micro size, but the most recent of these have begun to incorporate nanosized components.

Another way to produce nanosensors is through the bottom-up method, which involves assembling the sensors out of even more minuscule components, most likely individual atoms or molecules. This would involve moving atoms of a particular substance one by one into particular positions which, though it has been achieved in laboratory tests using tools such as atomic force microscopes, is still a significant difficulty, especially to do en masse, both for logistic reasons as well as economic ones. Most likely, this process would be used mainly for building starter molecules for self-assembling sensors.

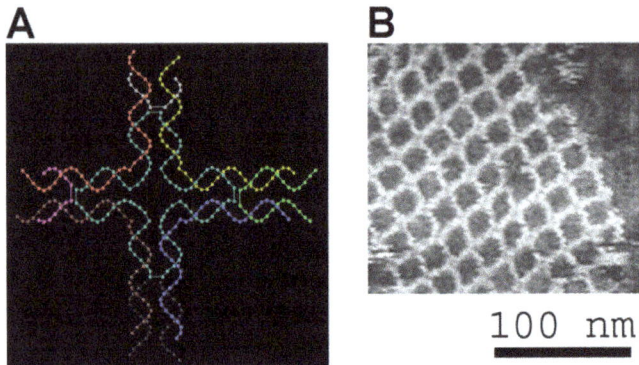

(A) An example of a DNA molecule used as a starter for larger self-assembly. (B) An atomic force microscope image of a self-assembled DNA nanogrid. Individual DNA tiles self-assemble into a highly ordered periodic two-dimensional DNA nanogrid.

The third way, which promises far faster results, involves self-assembly, or "growing" particular nanostructures to be used as sensors. This most often entails one of two types of assembly. The first involves using a piece of some previously created or naturally formed nanostructure and immersing it in free atoms of its own kind. After a given period, the structure, having an irregular surface that would make it prone to attracting more molecules as a continuation of its current pattern, would capture some of the free atoms and continue to form more of itself to make larger components of nanosensors.

The second type of self-assembly starts with an already complete set of components that would automatically assemble themselves into a finished product. Though this has been so far successful only in assembling computer chips at the micro size, researchers hope to eventually be able to do it at the nanometer size for multiple products, including nanosensors. Accurately being able to reproduce this effect for a desired sensor in a laboratory would imply that scientists could manufacture nanosensors much more quickly and potentially far more cheaply by letting numerous molecules assemble themselves with little or no outside influence, rather than having to manually assemble each sensor.

## Economic Impacts

Though nanosensor technology is a relatively new field, global projections for sales of products incorporating nanosensors range from $0.6 billion to $2.7 billion in the next three to four years. They will likely be included in most modern circuitry used in advanced computing systems, since their potential to provide the link between other forms of nanotechnology and the macroscopic world allows developers to fully exploit the potential of nanotechnology to miniaturize computer chips while vastly expanding their storage potential.

First, however, nanosensor developers must overcome the present high costs of production in order to become worthwhile for implementation in consumer products. Additionally, nanosensor reliability is not yet suitable for widespread use, and, because of their scarcity, nanosensors have yet to be marketed and implemented outside of research facilities. Consequently, nanosensors have yet to be made compatible with most consumer technologies for which they have been projected to eventually enhance.

## Social Impacts

Ethical and social impacts are harder to define and sort as good or bad compared to health and environmental impacts. The advancement in detecting and sensing different biological and chemical species with increased capacity and accuracy may transform societal mechanisms that were originally designed on uncertainty and imprecise information. For example, the ability to measure extremely low amounts of air pollutants or toxic materials in water raises questions and dilemmas of risk thresholds especially if the advancement of such technologies outpaces the ability of the public to respond. As another example, medical sensors will not only help in diagnoses and treatment but may also predict the future profile of an individual. This will add to the information used by health insurance companies to grant or deny coverage. Other social issues resulting from the widespread use of nanosensors and surveillance devices include privacy invasion and security issues.

## References

- Onoda, G.Y. Jr.; Hench, L.L., eds. (1979). Ceramic Processing Before Firing. New York: Wiley & Sons. ISBN 0-471-65410-8.

- MacNaught, Alan D.; Wilkinson, Andrew R., eds. (1997). Compendium of Chemical Terminology: IUPAC Recommendations (2nd ed.). Blackwell Science. ISBN 0865426848.

- Corriu, Robert & Anh, Nguyên Trong (2009). Molecular Chemistry of Sol-Gel Derived Nanomaterials. John Wiley and Sons. ISBN 0-470-72117-0.

- Pacios Pujadó, Mercè (2012). Carbon Nanotubes as Platforms for Biosensors with Electrochemical and Electronic Transduction. Springer Heidelberg. pp. XX,208. doi:10.1007/978-3-642-31421-6. ISBN 978-3-642-31421-6. ISSN 2190-5053.

- Osawa, Eiji (2002). Perspectives of Fullerene Nanotechnology. Springer Science & Business Media. pp. 29–. ISBN 978-0-7923-7174-8.

- Foster LE (2006). "Medical Nanotechnology: Science, Innovation, and Opportunity". Upper Saddle River: Pearson Education. ISBN 0-13-192756-6.

- atner MA; Ratner D; Ratner M. (2003). "Nanotechnology: A Gentle Introduction to the Next Big Idea". Upper Saddle River: Prentice Hall. ISBN 0-13-101400-5.

- Brinker, C.J. & Scherer, G.W. (1990). Sol-Gel Science: The Physics and Chemistry of Sol-Gel Processing. Academic Press. ISBN 0-12-134970-5.

- "General Safe Practices for Working with Engineered Nanomaterials in Research Laboratories". National Institute of Occupational Safety and Health. May 2012. pp. 6–8. Retrieved 2016-07-15.

- "Enhancing the efficiency of polymerase chain reaction using graphene nanoflakes - Abstract - Nanotechnology - IOPscience". iop.org. Retrieved 8 June 2015.

- "Nanofluid optical property characterization: towards efficient direct absorption solar collectors".

springer.com. Retrieved 8 June 2015.

- "Light: Science & Applications - Abstract of article: Nanofluid-based optical filter optimization for PV/T systems". nature.com. Retrieved 8 June 2015.

- Topmiller, Jennifer L.; Dunn, Kevin H. (9 December 2013). "Controlling Exposures to Workers Who Make or Use Nanomaterials". National Institute of Occupational Safety and Health. Retrieved 6 January 2015.

- "Current Strategies for Engineering Controls in Nanomaterial Production and Downstream Handling Processes" (PDF). National Institute of Occupational Safety and Health. November 2013. Retrieved 6 January 2015.

# 4

# Techniques of Nanoengineering

The techniques used for nanoengineering are scanning tunneling microscope and molecular self-assembly. Scanning tunneling microscope is an instrument that is used to scan surfaces at the atomic level whereas the process by which molecules arrange themselves without any guidance is termed as molecular self-assembly. This text discusses the methods of nanoengineering in a critical manner providing key analysis to the subject matter.

## Scanning Tunneling Microscope

A scanning tunneling microscope (STM) is an instrument for imaging surfaces at the atomic level. Its development in 1981 earned its inventors, Gerd Binnig and Heinrich Rohrer (at IBM Zürich), the Nobel Prize in Physics in 1986. For an STM, good resolution is considered to be 0.1 nm lateral resolution and 0.01 nm depth resolution. With this resolution, individual atoms within materials are routinely imaged and manipulated. The STM can be used not only in ultra-high vacuum but also in air, water, and various other liquid or gas ambients, and at temperatures ranging from near zero kelvin to a few hundred degrees Celsius.

The STM is based on the concept of quantum tunneling. When a conducting tip is brought very near to the surface to be examined, a bias (voltage difference) applied between the two can allow electrons to tunnel through the vacuum between them. The resulting *tunneling current* is a function of tip position, applied voltage, and the local density of states (LDOS) of the sample. Information is acquired by monitoring the current as the tip's position scans across the surface, and is usually displayed in image form. STM can be a challenging technique, as it requires extremely clean and stable surfaces, sharp tips, excellent vibration control, and sophisticated electronics, but nonetheless many hobbyists have built their own.

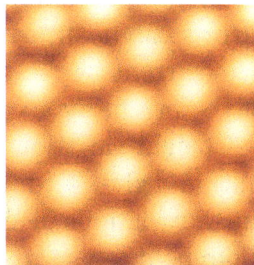

The silicon atoms on the surface of a crystal of silicon carbide (SiC). Image obtained using an STM.

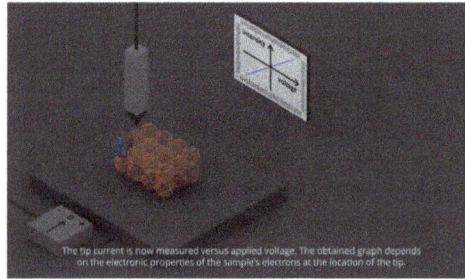

Scanning Tunneling Microscope operating principle

## Procedure

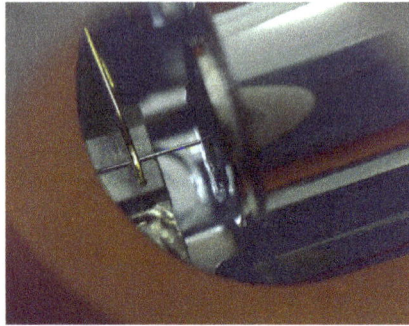

A close-up of a simple scanning tunneling microscope head using a platinum–iridium tip.

First, a voltage bias is applied and the tip is brought close to the sample by coarse sample-to-tip control, which is turned off when the tip and sample are sufficiently close. At close range, fine control of the tip in all three dimensions when near the sample is typically piezoelectric, maintaining tip-sample separation W typically in the 4-7 Å (0.4-0.7 nm) range, which is the equilibrium position between attractive (3<W<10Å) and repulsive (W<3Å) interactions. In this situation, the voltage bias will cause electrons to tunnel between the tip and sample, creating a current that can be measured. Once tunneling is established, the tip's bias and position with respect to the sample can be varied (with the details of this variation depending on the experiment) and data are obtained from the resulting changes in current.

If the tip is moved across the sample in the x-y plane, the changes in surface height and density of states cause changes in current. These changes are mapped in images. This change in current with respect to position can be measured itself, or the height, z, of the tip corresponding to a constant current can be measured. These two modes are called constant height mode and constant current mode, respectively. In constant current mode, feedback electronics adjust the height by a voltage to the piezoelectric height control mechanism. This leads to a height variation and thus the image comes from the tip topography across the sample and gives a constant charge density surface; this means contrast on the image is due to variations in charge density. In constant height mode, the voltage and height are both held constant while the current changes to keep the voltage from changing; this leads to an image made of current changes over the surface, which can be related to charge density. The benefit to using a constant height mode is that it is faster, as the piezoelectric movements require more time to register the height change in constant current mode than the current change in constant height mode. All images produced by STM are grayscale, with color optionally added in post-processing in order to visually emphasize important features.

In addition to scanning across the sample, information on the electronic structure at a given location in the sample can be obtained by sweeping voltage and measuring current at a specific location. This type of measurement is called scanning tunneling spectroscopy (STS) and typically results in a plot of the local density of states as a function of energy within the sample. The advantage of STM over other measurements of the density of states lies in its ability to make extremely local measurements: for example, the density of states at an impurity site can be compared to the density of states far from impurities.

Framerates of at least 25 Hz enable so called video-rate STM. Framerates up to 80 Hz are possible with fully working feedback that adjusts the height of the tip. Due to the line-by-line scanning motion, a proper comparison on the speed requires not only the framerate, but also the number of pixels in an image: with a framerate of 10Hz and 100x100 pixels the tip moves with a line frequency of 1 kHz, whereas it moves with only with 500 Hz, when measuring with a faster framerate of 50Hz but only 10x10 pixels. Video-rate STM can be used to scan surface diffusion.

## Instrumentation

The components of an STM include scanning tip, piezoelectric controlled height and x,y scanner, coarse sample-to-tip control, vibration isolation system, and computer.

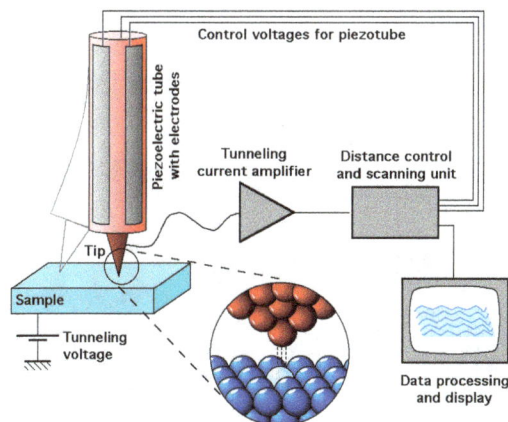

Schematic view of an STM

The resolution of an image is limited by the radius of curvature of the scanning tip of the STM. Additionally, image artifacts can occur if the tip has two tips at the end rather than a single atom; this leads to "double-tip imaging," a situation in which both tips contribute to the tunneling. Therefore, it has been essential to develop processes for consistently obtaining sharp, usable tips. Recently, carbon nanotubes have been used in this instance.

The tip is often made of tungsten or platinum-iridium, though gold is also used. Tungsten tips are usually made by electrochemical etching, and platinum-iridium tips by mechanical shearing.

Due to the extreme sensitivity of tunnel current to height, proper vibration insulation or an extremely rigid STM body is imperative for obtaining usable results. In the first STM by Binnig and Rohrer, magnetic levitation was used to keep the STM free from vibrations; now mechanical spring or gas spring systems are often used. Additionally, mechanisms for reducing eddy currents are sometimes implemented.

Maintaining the tip position with respect to the sample, scanning the sample and acquiring the data is computer controlled. The computer may also be used for enhancing the image with the help of image processing as well as performing quantitative measurements.

## Probe Tips

STM tips are usually made from tungsten metal or a platinum-iridium alloy where at the very end of the tip (called apex) there is one atom of the material.

## Other STM Related Studies

Many other microscopy techniques have been developed based upon STM. These include photon scanning microscopy (PSTM), which uses an optical tip to tunnel photons; scanning tunneling potentiometry (STP), which measures electric potential across a surface; spin polarized scanning tunneling microscopy (SPSTM), which uses a ferromagnetic tip to tunnel spin-polarized electrons into a magnetic sample, and atomic force microscopy (AFM), in which the force caused by interaction between the tip and sample is measured.

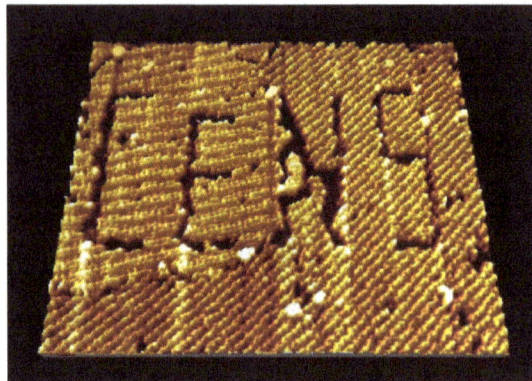

Nanomanipulation via STM of a self-assembled organic semiconductor monolayer (here: PTCDA molecules) on graphite, in which the logo of the Center for NanoScience (CeNS), LMU has been written.

Image of a graphite surface at an atomic level obtained by an STM.

Other STM methods involve manipulating the tip in order to change the topography of the sample. This is attractive for several reasons. Firstly the STM has an atomically precise positioning system which allows very accurate atomic scale manipulation. Furthermore, after the surface is modified by the tip, it is a simple matter to then image with the same tip, without changing the instrument. IBM researchers developed a way to manipulate xenon atoms adsorbed on a nickel surface. This technique has been used to create electron "corrals" with a small number of adsorbed atoms, which allows the STM to be used to observe electron Friedel oscillations on the surface of the material. Aside from modifying the actual sample surface, one can also use the STM to tunnel electrons into a layer of electron beam photoresist on a sample, in order to do lithography. This has the advantage of offering more control of the exposure than traditional electron beam lithography. Another practical application of STM is atomic deposition of metals (gold, silver, tungsten, etc.) with any desired (pre-programmed) pattern, which can be used as contacts to nanodevices or as nanodevices themselves.

Variable temperature STM was used to investigate temperature dependendy of molecular rotations on single crystalline surfaces. Rotating molecules appear blurred compared to non-rotating ones.

Recently groups have found they can use the STM tip to rotate individual bonds within single molecules. The electrical resistance of the molecule depends on the orientation of the bond, so the molecule effectively becomes a molecular switch.

## Principle of Operation

The first STM produced commercially, 1986.

Tunneling is a functioning concept that arises from quantum mechanics. Classically, an object hitting an impenetrable barrier will not pass through. In contrast, objects with a very small mass, such as the electron, have wavelike characteristics which permit such an event, referred to as tunneling.

Electrons behave as beams of energy, and in the presence of a potential $U(z)$, assuming 1-dimensional case, the energy levels $\psi_n(z)$ of the electrons are given by solutions to Schrödinger's equation,

$$-\frac{\hbar^2}{2m}\frac{\partial^2 \psi_n(z)}{\partial z^2}+U(z)\psi_n(z)=E\psi_n(z)$$

where $\hbar$ is the reduced Planck's constant, $z$ is the position, and $m$ is the mass of an electron. If an electron of energy $E$ is incident upon an energy barrier of height $U(z)$, the electron wave function is a traveling wave solution,

$$\psi_n(z)=\psi_n(0)e^{\pm ikz}$$

where

$$k=\frac{\sqrt{2m(E-U(z))}}{\hbar}$$

if $E > U(z)$, which is true for a wave function inside the tip or inside the sample. Inside a barrier, $E < U(z)$ so the wave functions which satisfy this are decaying waves,

$$\psi_n(z)=\psi_n(0)e^{\pm \kappa z}$$

where

$$\kappa=\frac{\sqrt{2m(U-E)}}{\hbar}$$

quantifies the decay of the wave inside the barrier, with the barrier in the $+z$ direction for $-\kappa$. Knowing the wave function allows one to calculate the probability density for that electron to be found at some location. In the case of tunneling, the tip and sample wave functions overlap such that when under a bias, there is some finite probability to find the electron in the barrier region and even on the other side of the barrier. Let us assume the bias is $V$ and the barrier width is $W$. This probability, $P$, that an electron at $z=0$ (left edge of barrier) can be found at $z=W$ (right edge of barrier) is proportional to the wave function squared,

$$P \propto |\psi_n(0)|^2\, e^{-2\kappa W}.$$

If the bias is small, we can let $U - E \approx \varphi M$ in the expression for $\kappa$, where $\varphi M$, the work function, gives the minimum energy needed to bring an electron from an occupied level, the highest of which is at the Fermi level (for metals at $T=0$ kelvins), to vacuum level. When a small bias $V$ is applied to the system, only electronic states very near the Fermi level, within $eV$ (a product of electron charge and voltage, not to be confused here with electronvolt unit), are excited. These excited electrons can tunnel across the barrier. In other words, tunneling occurs mainly with electrons of energies near the Fermi level.

A large scanning tunneling microscope, in the labs of the London Centre for Nanotechnology

However, tunneling does require that there is an empty level of the same energy as the electron for the electron to tunnel into on the other side of the barrier. It is because of this restriction that the tunneling current can be related to the density of available or filled states in the sample. The current due to an applied voltage $V$ (assume tunneling occurs sample to tip) depends on two factors: 1) the number of electrons between $E_f$ and $eV$ in the sample, and 2) the number among them which have corresponding free states to tunnel into on the other side of the barrier at the tip. The higher the density of available states the greater the tunneling current. When $V$ is positive, electrons in the tip tunnel into empty states in the sample; for a negative bias, electrons tunnel out of occupied states in the sample into the tip.

Mathematically, this tunneling current is given by

$$I \propto \sum_{E_f - eV}^{E_f} |\psi_n(0)|^2 \, e^{-2\kappa W}.$$

One can sum the probability over energies between $E_f - eV$ and $E_f$ to get the number of states available in this energy range per unit volume, thereby finding the local density of states (LDOS) near the Fermi level. The LDOS near some energy $E$ in an interval $\varepsilon$ is given by

$$\rho_s(z, E) = \frac{1}{\varepsilon} \sum_{E - \varepsilon}^{E} |\psi_n(z)|^2,$$

and the tunnel current at a small bias V is proportional to the LDOS near the Fermi level, which gives important information about the sample. It is desirable to use LDOS to express the current because this value does not change as the volume changes, while probability density does. Thus the tunneling current is given by

$$I \propto V \rho_s(0, E_f) e^{-2\kappa W}$$

where $\rho_s(0, E_f)$ is the LDOS near the Fermi level of the sample at the sample surface. This current can also be expressed in terms of the LDOS near the Fermi level of the sample at the tip surface,

$$I \propto V \rho_s(W, E_f)$$

The exponential term in the above equations means that small variations in W greatly influence the tunnel current. If the separation is decreased by 1 Å, the current increases by an order of magnitude, and vice versa.

This approach fails to account for the *rate* at which electrons can pass the barrier. This rate should affect the tunnel current, so it can be treated using the Fermi's golden rule with the appropriate tunneling matrix element. John Bardeen solved this problem in his study of the metal-insulator-metal junction. He found that if he solved Schrödinger's equation for each side of the junction separately to obtain the wave functions ψ and χ for each electrode, he could obtain the tunnel matrix, M, from the overlap of these two wave functions. This can be applied to STM by making the electrodes the tip and sample, assigning ψ and χ as sample and tip wave functions, respectively, and evaluating M at some surface S between the metal electrodes, where z=0 at the sample surface and z=W at the tip surface.

Now, Fermi's Golden Rule gives the rate for electron transfer across the barrier, and is written

$$w = \frac{2\pi}{\hbar} |M|^2 \delta(E_\psi - E_\chi),$$

where $\delta(E_\psi - E_\chi)$ restricts tunneling to occur only between electron levels with the same energy. The tunnel matrix element, given by

$$M = \frac{\hbar^2}{2m} \int_{z=z_0} (\chi^* \frac{\partial \psi}{\partial z} - \psi \frac{\partial \chi^*}{\partial z}) dS,$$

is a description of the lower energy associated with the interaction of wave functions at the overlap, also called the resonance energy.

Summing over all the states gives the tunneling current as

$$I = \frac{4\pi e}{\hbar} \int_{-\infty}^{+\infty} [f(E_f - eV + \epsilon) - f(E_f + \epsilon)] \rho_s (E_f - eV + \epsilon) \rho_T (E_f + \epsilon) |M|^2 d\epsilon$$

where $f$ is the Fermi function, $\rho_s$ and $\rho_T$ are the density of states in the sample and tip, respectively. The Fermi distribution function describes the filling of electron levels at a given temperature T.

## Early Invention

An earlier, similar invention, the *Topografiner* of R. Young, J. Ward, and F. Scire from the NIST, relied on field emission. However, Young is credited by the Nobel Committee as the person who realized that it should be possible to achieve better resolution by using the tunnel effect.

# Molecular Self-assembly

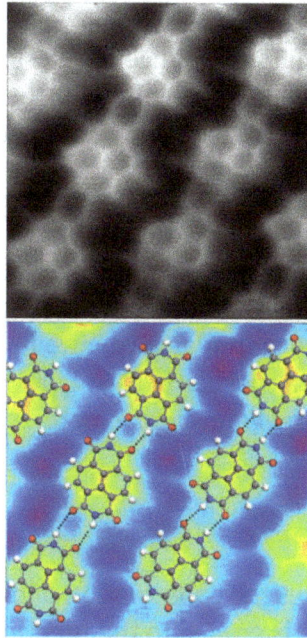

AFM image of napthalenetetracarboxylic diimide molecules on silver interacting via hydrogen bonding (77 K).

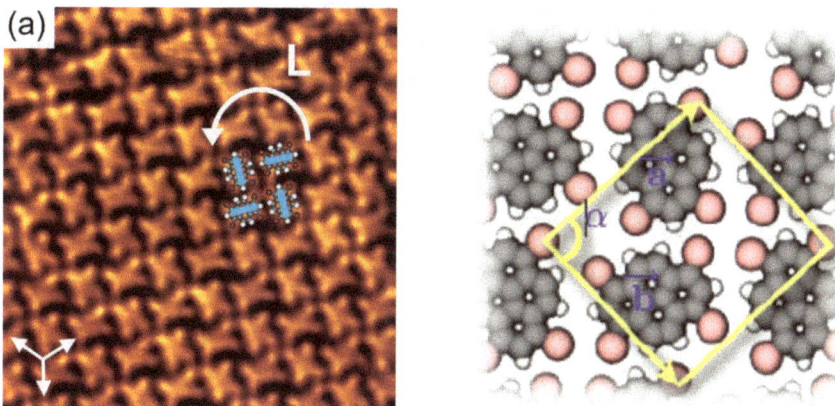

STM image of self-assembled Br$_4$-pyrene molecules on Au(111) surface (top) and its model (bottom; pink spheres are Br atoms).

Molecular self-assembly is the process by which molecules adopt a defined arrangement without guidance or management from an outside source. There are two types of self-assembly. These are intramolecular self-assembly and intermolecular self-assembly. Commonly, the term molecular self-assembly refers to intermolecular self-assembly, while the intramolecular analog is more commonly called folding.

## Supramolecular Systems

Molecular self-assembly is a key concept in supramolecular chemistry. This is because assembly of molecules in such systems is directed through noncovalent interactions (e.g., hydrogen bonding, metal coordination, hydrophobic forces, van der Waals forces, π-π interactions, and/or electrostatic) as well as electromagnetic interactions. Common examples include the formation of micelles, vesicles, liquid crystal phases, and Langmuir monolayers by surfactant molecules. Further examples of supramolecular assemblies demonstrate that a variety of different shapes and sizes can be obtained using molecular self-assembly.

Molecular self-assembly allows the construction of challenging molecular topologies. One example is Borromean rings, interlocking rings wherein removal of one ring unlocks each of the other rings. DNA has been used to prepare a molecular analog of Borromean rings. More recently, a similar structure has been prepared using non-biological building blocks. While a mechanistic understanding of how supramolecular self-assembly occurs remains largely unknown, both experimental and theoretic work has been pursued on this topic.

## Biological Systems

Molecular self-assembly underlies the construction of biologic macromolecular assemblies in living organisms, and so is crucial to the function of cells. It is exhibited in the self-assembly of lipids to form the membrane, the formation of double helical DNA through hydrogen bonding of the individual strands, and the assembly of proteins to form quaternary structures. Molecular self-assembly of incorrectly folded proteins into insoluble amyloid fibers is responsible for infectious prion-related neurodegenerative diseases. Molecular self-assembly of nanoscale structures plays a role in the growth of the remarkable β-keratin lamellae/setae/spatulae structures used to give geckos the ability to climb walls and adhere to ceilings and rock overhangs.

## Nanotechnology

Molecular self-assembly is an important aspect of bottom-up approaches to nanotechnology. Using molecular self-assembly the final (desired) structure is programmed in the shape and functional groups of the molecules. Self-assembly is referred to as a 'bottom-up' manufacturing technique in contrast to a 'top-down' technique such as lithography where the desired final structure is carved from a larger block of matter. In the speculative vision of molecular nanotechnology, microchips of the future might be made by molecular self-assembly. An advantage to constructing nanostructure using molecular self-assembly for biological materials is that they will degrade back into individual molecules that can be broken down by the body.

## DNA Nanotechnology

DNA nanotechnology is an area of current research that uses the bottom-up, self-assembly approach for nanotechnological goals. DNA nanotechnology uses the unique molecular recognition properties of DNA and other nucleic acids to create self-assembling branched DNA complexes with useful properties. DNA is thus used as a structural material rather than as a carrier of biological information, to make structures such as two-dimensional periodic lattices (both tile-based as well as using the "DNA origami" method) and three-dimensional structures in the shapes of polyhedra. These DNA structures have also been used as templates in the assembly of other molecules such as gold nanoparticles and streptavidin proteins.

## Two-dimensional Monolayers

The spontaneous assembly of a single layer of molecules at interfaces is usually referred to as two-dimensional self-assembly. Early direct proofs showing that molecules can assembly into higher-order architectures at solid interfaces came with the development of scanning tunneling microscopy and shortly thereafter. Eventually two strategies became popular for the self-assembly of 2D architectures, namely self-assembly following ultra-high-vacuum deposition and annealing and self-assembly at the solid-liquid interface. The design of molecules and conditions leading to the formation of highly-crystalline architectures is considered today a form of 2D crystal engineering at the nanoscopic scale.

## References

- K. Oura; V. G. Lifshits; A. A. Saranin; A. V. Zotov & M. Katayama (2003). Surface science: an introduction. Berlin: Springer-Verlag. ISBN 3-540-00545-5.

- D. A. Bonnell & B. D. Huey (2001). "Basic principles of scanning probe microscopy". In D. A. Bonnell. Scanning probe microscopy and spectroscopy: Theory, techniques, and applications (2 ed.). New York: Wiley-VCH. ISBN 0-471-24824-X.

- R. V. Lapshin (2011). "Feature-oriented scanning probe microscopy". In H. S. Nalwa. Encyclopedia of Nanoscience and Nanotechnology (PDF). 14. USA: American Scientific Publishers. pp. 105–115. ISBN 1-58883-163-9.

- Rosen, Milton J. (2004). Surfactants and interfacial phenomena. Hoboken, NJ: Wiley-Interscience. ISBN 978-0-471-47818-8.

- C. Bai (2000). Scanning tunneling microscopy and its applications. New York: Springer Verlag. ISBN 3-540-65715-0.

- C. Julian Chen (1993). Introduction to Scanning Tunneling Microscopy (PDF). Oxford University Press. ISBN 0-19-507150-6.

- M. J. Rost; et al. (August 2016). "Scanning probe microscopes go video rate and beyond". Review of Scientific Instruments. US: American Institute of Physics. 76: 053710. doi:10.1063/1.1915288. ISSN 1369-7021.

- Pham, Tuan Anh; Song, Fei; Nguyen, Manh-Thuong; Stöhr, Meike (2014). "Self-assembly of pyrene derivatives on Au(111): Substituent effects on intermolecular interactions". Chem. Commun. 50 (91): 14089. doi:10.1039/C4CC02753A.

- T. Waldmann; J. Klein; H.E. Hoster; R.J. Behm (2012). "Stabilization of Large Adsorbates by Rotational Entropy: A Time-Resolved Variable-Temperature STM Study". ChemPhysChem. 13: 162–169. doi:10.1002/cphc.201200531.

# 5

# Applications and Designs of Nanoengineering

This section discusses the applications and designs of nanoengineering. The medical function of nanotechnology is nanomedicine whereas the study of nanomaterials is nanotoxicology. The other applications and designs that have been discussed are nano-electronics, green nanotechnology, DNA nanotechnology, nanoremediation and nano-filtration. The topics elaborated in the chapter will help in gaining a better perspective about the applications of nanoengineering.

## Nanomedicine

Nanomedicine is the medical application of nanotechnology. Nanomedicine ranges from the medical applications of nanomaterials and biological devices, to nanoelectronic biosensors, and even possible future applications of molecular nanotechnology such as biological machines. Current problems for nanomedicine involve understanding the issues related to toxicity and environmental impact of nanoscale materials (materials whose structure is on the scale of nanometers, i.e. billionths of a meter).

Functionalities can be added to nanomaterials by interfacing them with biological molecules or structures. The size of nanomaterials is similar to that of most biological molecules and structures; therefore, nanomaterials can be useful for both in vivo and in vitro biomedical research and applications. Thus far, the integration of nanomaterials with biology has led to the development of diagnostic devices, contrast agents, analytical tools, physical therapy applications, and drug delivery vehicles.

Nanomedicine seeks to deliver a valuable set of research tools and clinically useful devices in the near future. The National Nanotechnology Initiative expects new commercial applications in the pharmaceutical industry that may include advanced drug delivery systems, new therapies, and in vivo imaging. Nanomedicine research is receiving funding from the US National Institutes of Health, including the funding in 2005 of a five-year plan to set up four nanomedicine centers.

Nanomedicine sales reached $16 billion in 2015, with a minimum of $3.8 billion in nanotechnology R&D being invested every year. Global funding for emerging nanotechnology increased by 45% per year in recent years, with product sales exceeding $1 trillion in 2013. As the nanomedicine industry continues to grow, it is expected to have a significant impact on the economy.

## Drug Delivery

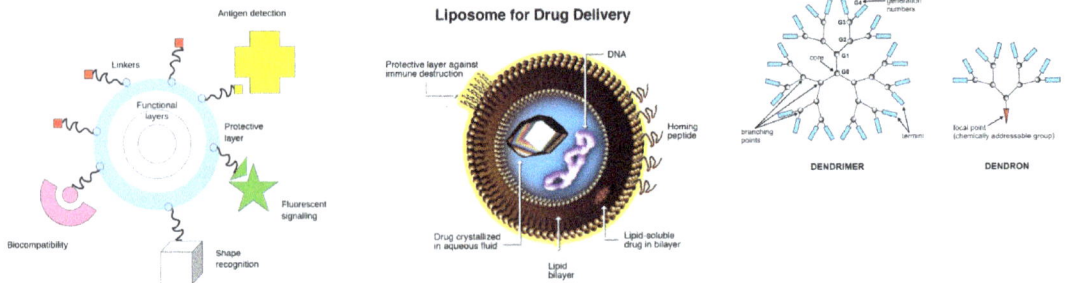

Nanoparticles *(top)*, liposomes *(middle)*, and dendrimers *(bottom)* are some nanomaterials being investigated for use in nanomedicine.

Nanotechnology has provided the possibility of delivering drugs to specific cells using nanoparticles.

The overall drug consumption and side-effects may be lowered significantly by depositing the active agent in the morbid region only and in no higher dose than needed. Targeted drug delivery is intended to reduce the side effects of drugs with concomitant decreases in consumption and treatment expenses. Drug delivery focuses on maximizing bioavail-ability both at specific places in the body and over a period of time. This can potentially be achieved by molecular targeting by nanoengineered devices. More than $65 billion are wasted each year due to poor bioavailability. A benefit of using nanoscale for medi-cal technologies is that smaller devices are less invasive and can possibly be implanted inside the body, plus biochemical reaction times are much shorter. These devices are faster and more sensitive than typical drug delivery. The efficacy of drug delivery through nanomedicine is largely based upon: a) efficient encapsulation of the drugs, b) successful delivery of drug to the targeted region of the body, and c) successful release of the drug.

Drug delivery systems, lipid- or polymer-based nanoparticles, can be designed to im-prove the pharmacokinetics and biodistribution of the drug. However, the pharmaco-kinetics and pharmacodynamics of nanomedicine is highly variable among different patients. When designed to avoid the body's defence mechanisms, nanoparticles have beneficial properties that can be used to improve drug delivery. Complex drug delivery mechanisms are being developed, including the ability to get drugs through cell mem-branes and into cell cytoplasm. Triggered response is one way for drug molecules to be used more efficiently. Drugs are placed in the body and only activate on encountering a particular signal. For example, a drug with poor solubility will be replaced by a drug delivery system where both hydrophilic and hydrophobic environments exist, improv-ing the solubility. Drug delivery systems may also be able to prevent tissue damage through regulated drug release; reduce drug clearance rates; or lower the volume of distribution and reduce the effect on non-target tissue. However, the biodistribution of these nanoparticles is still imperfect due to the complex host's reactions to nano- and microsized materials and the difficulty in targeting specific organs in the body. Nevertheless, a lot of work is still ongoing to optimize and better understand the potential and limitations of nanoparticulate systems. While advancement of research proves that targeting and distribution can be augmented by nanoparticles, the dangers of nano-toxicity become an important next step in further understanding of their medical uses.

Nanoparticles can be used in combination therapy for decreasing antibiotic resistance or for their antimicrobial properties. Nanoparticles might also used to circumvent multidrug resistance (MDR) mechanisms.

## Types of Systems used

Two forms of nanomedicine that have already been tested in mice and are awaiting human trials that will be using gold nanoshells to help diagnose and treat cancer, and using liposomes as vaccine adjuvants and as vehicles for drug transport. Similarly, drug detoxification is also another application for nanomedicine which has shown promising results in rats. Advances in Lipid nanotechnology was also instrumental in engineering medical nanodevices and novel drug delivery systems as well as in developing sensing applications. Another example can be found in dendrimers and nanoporous materials. Another example is to use block co-polymers, which form micelles for drug encapsulation.

Polymeric nano-particles are a competing technology to lipidic (based mainly on Phospholipids) nano-particles. There is an additional risk of toxicity associated with polymers not widely studied or understood. The major advantages of polymers is stability, lower cost and predictable characterisation. However, in the patient's body this very stability (slow degradation) is a negative factor. Phospholipids on the other hand are membrane lipids (already present in the body and surrounding each cell), have a GRAS (Generally Recognised As Safe) status from FDA and are derived from natural sources without any complex chemistry involved. They are not metabolised but rather absorbed by the body and the degradation products are themselves nutrients (fats or micronutrients).

Protein and peptides exert multiple biological actions in the human body and they have been identified as showing great promise for treatment of various diseases and disorders. These macromolecules are called biopharmaceuticals. Targeted and/or controlled delivery of these biopharmaceuticals using nanomaterials like nanoparticles and Dendrimers is an emerging field called nanobiopharmaceutics, and these products are called nanobiopharmaceuticals

Another highly efficient system for microRNA delivery for example are nanoparticles formed by the self-assembly of two different microRNAs deregulated in cancer.

Another vision is based on small electromechanical systems; nanoelectromechanical systems are being investigated for the active release of drugs. Some potentially important applications include cancer treatment with iron nanoparticles or gold shells. Nanotechnology is also opening up new opportunities in implantable delivery systems, which are often preferable to the use of injectable drugs, because the latter frequently display first-order kinetics (the blood concentration goes up rapidly, but drops exponentially over time). This rapid rise may cause difficulties with toxicity, and drug efficacy can diminish as the drug concentration falls below the targeted range.

## Applications

Some nanotechnology-based drugs that are commercially available or in human clinical trials include:

- Abraxane, approved by the U.S. Food and Drug Administration (FDA) to treat breast cancer, non-small- cell lung cancer (NSCLC) and pancreatic cancer, is the nanoparticle albumin bound paclitaxel.

- Doxil was originally approved by the FDA for the use on HIV-related Kaposi's sarcoma. It is now being used to also treat ovarian cancer and multiple myeloma. The drug is encased in liposomes, which helps to extend the life of the drug that is being distributed. Liposomes are self-assembling, spherical, closed colloidal structures that are composed of lipid bilayers that surround an aqueous space. The liposomes also help to increase the functionality and it helps to decrease the damage that the drug does to the heart muscles specifically.

- Onivyde, liposome encapsulated irinotecan to treat metastatic pancreatic cancer, was approved by FDA in October 2015.

- C-dots (Cornell dots) are the smallest silica-based nanoparticles with the size <10 nm. The particles are infused with organic dye which will light up with fluorescence. Clinical trial is underway since 2011 to use the C-dots as diagnostic tool to assist surgeons to identify the location of tumor cells.

- An early phase clinical trial using the platform of 'Minicell' nanoparticle for drug delivery have been tested on patients with advanced and untreatable cancer. Built from the membranes of mutant bacteria, the minicells were loaded with paclitaxel and coated with cetuximab, antibodies that bind the epidermal growth factor receptor (EGFR) which is often overexpressed in a number of cancers, as a 'homing' device to the tumor cells. The tumor cells recognize the bacteria from which the minicells have been derived, regard it as invading microorganism and engulf it. Once inside, the payload of anti-cancer drug kills the tumor cells. Measured at 400 nanometers, the minicell is bigger than synthetic particles developed for drug delivery. The researchers indicated that this larger size gives the minicells a better profile in side-effects because the minicells will preferentially leak out of the porous blood vessels around the tumor cells and do not reach the liver, digestive system and skin. This Phase 1 clinical trial demonstrated that this treatment is well tolerated by the patients. As a platform technology, the minicell drug delivery system can be used to treat a number of different cancers with different anti-cancer drugs with the benefit of lower dose and less side-effects.

- In 2014, a Phase 3 clinical trial for treating inflammation and pain after cataract surgery, and a Phase 2 trial for treating dry eye disease were initiated using nanoparticle loteprednol etabonate. In 2015, the product, KPI-121 was found to produce statistically significant positive results for the post-surgery treatment.

## Cancer

Existing and potential drug nanocarriers have been reviewed.

Nanoparticles have high surface area to volume ratio. This allows for many functional groups to be attached to a nanoparticle, which can seek out and bind to certain tumor cells. Additionally, the small size of nanoparticles (10 to 100 nanometers), allows them to preferentially accumulate at tumor sites (because tumors lack an effective lymphatic drainage system). Limitations to conventional cancer chemotherapy include drug resistance, lack of selectivity, and lack of solubility. Nanoparticles have the potential to overcome these problems.

In photodynamic therapy, a particle is placed within the body and is illuminated with light from the outside. The light gets absorbed by the particle and if the particle is metal, energy from the light will heat the particle and surrounding tissue. Light may also be used to produce high energy oxygen molecules which will chemically react with and destroy most organic molecules that are next to them (like tumors). This therapy is appealing for many reasons. It does not leave a "toxic trail" of reactive molecules throughout the body (chemotherapy) because it is directed where only the light is shined and the particles exist. Photodynamic therapy has potential for a noninvasive procedure for dealing with diseases, growth and tumors. Kanzius RF therapy is one example of such therapy (nanoparticle hyperthermia). Also, gold nanoparticles have the potential to join numerous therapeutic functions into a single platform, by targeting specific tumor cells, tissues and organs.

## Visualization

*In vivo* imaging is another area where tools and devices are being developed. Using nanoparticle contrast agents, images such as ultrasound and MRI have a favorable distribution and improved contrast. This might be accomplished by self assembled biocompatible nanodevices that will detect, evaluate, treat and report to the clinical doctor automatically.

The small size of nanoparticles endows them with properties that can be very useful in oncology, particularly in imaging. Quantum dots (nanoparticles with quantum confinement properties, such as size-tunable light emission), when used in conjunction with MRI (magnetic resonance imaging), can produce exceptional images of tumor sites. Nanoparticles of cadmium selenide (quantum dots) glow when exposed to ultraviolet light. When injected, they seep into cancer tumors. The surgeon can see the glowing tumor, and use it as a guide for more accurate tumor removal. These nanoparticles are much brighter than organic dyes and only need one light source for excitation. This means that the use of fluorescent quantum dots could produce a higher contrast image and at a lower cost than today's organic dyes used as contrast media. The downside, however, is that quantum dots are usually made of quite toxic elements.

Tracking movement can help determine how well drugs are being distributed or how substances are metabolized. It is difficult to track a small group of cells throughout the body, so scientists used to dye the cells. These dyes needed to be excited by light of a certain wavelength in order for them to light up. While different color dyes absorb different frequencies of light, there was a need for as many light sources as cells. A way around this problem is with luminescent tags. These tags are quantum dots attached to proteins that penetrate cell membranes. The dots can be random in size, can be made of bio-inert material, and they demonstrate the nanoscale property that color is size-dependent. As a result, sizes are selected so that the frequency of light used to make a group of quantum dots fluoresce is an even multiple of the frequency required to make another group incandesce. Then both groups can be lit with a single light source. They have also found a way to insert nanoparticles into the affected parts of the body so that those parts of the body will glow showing the tumor growth or shrinkage or also organ trouble.

## Sensing

Nanotechnology-on-a-chip is one more dimension of lab-on-a-chip technology. Magnetic nanoparticles, bound to a suitable antibody, are used to label specific molecules, structures or microorganisms. Gold nanoparticles tagged with short segments of DNA can be used for detection of genetic sequence in a sample. Multicolor optical coding for biological assays has been achieved by embedding different-sized quantum dots into polymeric microbeads. Nanopore technology for analysis of nucleic acids converts strings of nucleotides directly into electronic signatures.

Sensor test chips containing thousands of nanowires, able to detect proteins and other biomarkers left behind by cancer cells, could enable the detection and diagnosis of cancer in the early stages from a few drops of a patient's blood. Nanotechnology is helping to advance the use of arthroscopes, which are pencil-sized devices that are used in surgeries with lights and cameras so surgeons can do the surgeries with smaller incisions. The smaller the incisions the faster the healing time which is better for the patients. It is also helping to find a way to make an arthroscope smaller than a strand of hair.

Research on nanoelectronics-based cancer diagnostics could lead to tests that can be done in pharmacies. The results promise to be highly accurate and the product promises to be inexpensive. They could take a very small amount of blood and detect cancer anywhere in the body in about five minutes, with a sensitivity that is a thousand times better than in a conventional laboratory test. These devices that are built with nanowires to detect cancer proteins; each nanowire detector is primed to be sensitive to a different cancer marker. The biggest advantage of the nanowire detectors is that they could test for anywhere from ten to one hundred similar medical conditions without adding cost to the testing device. Nanotechnology has also helped to personalize oncology for the detection, diagnosis, and treatment of cancer. It is now able to be tailored to each individual's tumor for better performance. They have found ways that they will be able to target a specific part of the body that is being affected by cancer.

## Blood Purification

Magnetic micro particles are proven research instruments for the separation of cells and proteins from complex media. The technology is available under the name Magnetic-activated cell sorting or Dynabeads among others. More recently it was shown in animal models that magnetic nanoparticles can be used for the removal of various noxious compounds including toxins, pathogens, and proteins from whole blood in an extracorporeal circuit similar to dialysis. In contrast to dialysis, which works on the principle of the size related diffusion of solutes and ultrafiltration of fluid across a semi-permeable membrane, the purification with nanoparticles allows specific targeting of substances. Additionally larger compounds which are commonly not dialyzable can be removed.

The purification process is based on functionalized iron oxide or carbon coated metal nanoparticles with ferromagnetic or superparamagnetic properties. Binding agents such as proteins, antibodies, antibiotics, or synthetic ligands are covalently linked to the particle surface. These binding agents are able to interact with target species forming an agglomerate. Applying an external magnetic field gradient allows exerting a force on the nanoparticles. Hence the particles can be separated from the bulk fluid, thereby cleaning it from the contaminants.

The small size (< 100 nm) and large surface area of functionalized nanomagnets leads to advantageous properties compared to hemoperfusion, which is a clinically used technique for the purification of blood and is based on surface adsorption. These advantages are high loading and accessibility of the binding agents, high selectivity towards the target compound, fast diffusion, small hydrodynamic resistance, and low dosage.

This approach offers new therapeutic possibilities for the treatment of systemic infections such as sepsis by directly removing the pathogen. It can also be used to selectively remove cytokines or endotoxins or for the dialysis of compounds which are not accessible by traditional dialysis methods. However the technology is still in a preclinical phase and first clinical trials are not expected before 2017.

## Tissue Engineering

Nanotechnology may be used as part of tissue engineering to help reproduce or repair or reshape damaged tissue using suitable nanomaterial-based scaffolds and growth factors. Tissue engineering if successful may replace conventional treatments like organ transplants or artificial implants. Nanoparticles such as graphene, carbon nanotubes, molybdenum disulfide and tungsten disulfide are being used as reinforcing agents to fabricate mechanically strong biodegradable polymeric nanocomposites for bone tissue engineering applications. The addition of these nanoparticles in the polymer matrix at low concentrations (~0.2 weight %) leads to significant improvements in the compressive and flexural mechanical properties of polymeric nanocomposites. Potentially, these nanocomposites may be used as a novel, mechanically strong, light weight composite as bone implants.

For example, a flesh welder was demonstrated to fuse two pieces of chicken meat into a single piece using a suspension of gold-coated nanoshells activated by an infrared laser. This could be used to weld arteries during surgery. Another example is nanonephrology, the use of nanomedicine on the kidney.

## Medical Devices

Neuro-electronic interfacing is a visionary goal dealing with the construction of nanodevices that will permit computers to be joined and linked to the nervous system. This idea requires the building of a molecular structure that will permit control and detection of nerve impulses by an external computer. A refuelable strategy implies energy is refilled continuously or periodically with external sonic, chemical, tethered, magnetic, or biological electrical sources, while a nonrefuelable strategy implies that all power is drawn from internal energy storage which would stop when all energy is drained. A nanoscale enzymatic biofuel cell for self-powered nanodevices have been developed that uses glucose from biofluids including human blood and watermelons. One limitation to this innovation is the fact that electrical interference or leakage or overheating from power consumption is possible. The wiring of the structure is extremely difficult because they must be positioned precisely in the nervous system. The structures that will provide the interface must also be compatible with the body's immune system.

Molecular nanotechnology is a speculative subfield of nanotechnology regarding the possibility of engineering molecular assemblers, machines which could re-order matter at a molecular or atomic scale. Nanomedicine would make use of these nanorobots, introduced into the body, to repair or detect damages and infections. Molecular nanotechnology is highly theoretical, seeking to anticipate what inventions nanotechnology might yield and to propose an agenda for future inquiry. The proposed elements of molecular nanotechnology, such as molecular assemblers and nanorobots are far beyond current capabilities. Future advances in nanomedicine could give rise to life extension through the repair of many processes thought to be responsible for aging. K. Eric Drexler, one of the founders of nanotechnology, postulated cell repair machines, including ones operating within cells and utilizing as yet hypothetical molecular machines, in his 1986 book Engines of Creation, with the first technical discussion of medical nanorobots by Robert Freitas appearing in 1999. Raymond Kurzweil, a futurist and transhumanist, stated in his book *The Singularity Is Near* that he believes that advanced medical nanorobotics could completely remedy the effects of aging by 2030. According to Richard Feynman, it was his former graduate student and collaborator Albert Hibbs who originally suggested to him (circa 1959) the idea of a *medical* use for Feynman's theoretical micromachines. Hibbs suggested that certain repair machines might one day be reduced in size to the point that it would, in theory, be possible to (as Feynman put it) "swallow the doctor". The idea was incorporated into Feynman's 1959 essay *There's Plenty of Room at the Bot-tom.*

# Nanotoxicology

Nanotoxicology is the study of the toxicity of nanomaterials. Because of quantum size effects and large surface area to volume ratio, nanomaterials have unique properties compared with their larger counterparts.

Nanotoxicology is a branch of bionanoscience which deals with the study and application of toxicity of nanomaterials. Nanomaterials, even when made of inert elements like gold, become highly active at nanometer dimensions. Nanotoxicological studies are intended to determine whether and to what extent these properties may pose a threat to the environment and to human beings. For instance, Diesel nanoparticles have been found to damage the cardiovascular system in a mouse model.

## Background

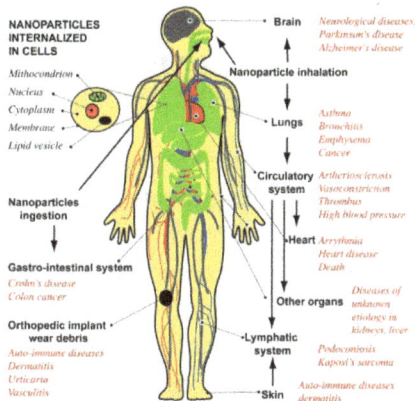

Pathways of exposure to nanoparticles and associated diseases as suggested by epidemiological, in vivo and in vitro studies.

Nanotoxicology is a sub-specialty of particle toxicology. It addresses the toxicology of nanoparticles (particles <100 nm diameter) which appear to have toxicity effects that are unusual and not seen with larger particles. Nanoparticles can be divided into combustion-derived nanoparticles (like diesel soot), manufactured nanoparticles like carbon nanotubes and naturally occurring nanoparticles from volcanic eruptions, atmospheric chemistry etc. Typical nanoparticles that have been studied are titanium dioxide, alumina, zinc oxide, carbon black, and carbon nanotubes, and "nano-$C_{60}$". Nanoparticles have much larger surface area to unit mass ratios which in some cases may lead to greater pro-inflammatory effects (in, for example, lung tissue). In addition, some nanoparticles seem to be able to translocate from their site of deposition to distant sites such as the blood and the brain. This has resulted in a sea-change in how particle toxicology is viewed- instead of being confined to the lungs, nanoparticle toxicologists study the brain, blood, liver, skin and gut.

Calls for tighter regulation of nanotechnology have arisen alongside a growing debate related to the human health and safety risks associated with nanotechnology. From a large-scale literature review, Yaobo Ding et al. found that release of airborne engineered nanoparticles and associated worker exposure from various production and handling activities at different workplaces are very probable. The Royal Society identifies the potential for nanoparticles to penetrate the skin, and recommends that the use of nanoparticles in cosmetics be conditional upon a favorable assessment by the relevant European Commission safety advisory committee. Andrew Maynard also reports that 'certain nanoparticles may move easily into sensitive lung tissues after inhalation, and cause damage that can lead to chronic breathing problems'.

Carbon nanotubes – characterized by their microscopic size and incredible tensile strength – are frequently likened to asbestos, due to their needle-like fiber shape. In a recent study that introduced carbon nanotubes into the abdominal cavity of mice, results demonstrated that long thin carbon nanotubes showed the same effects as long thin asbestos fibers, raising concerns that exposure to carbon nanotubes may lead to pleural abnormalities such as mesothelioma (cancer of the lining of the lungs caused by exposure to asbestos). Given these risks, effective and rigorous regulation has been called for to determine if, and under what circumstances, carbon nanotubes are manufactured, as well as ensuring their safe handling and disposal.

The Woodrow Wilson Centre's Project on Emerging Technologies conclude that there is insufficient funding for human health and safety research, and as a result there is currently limited understanding of the human health and safety risks associated with nanotechnology. While the US National Nanotechnology Initiative reports that around four percent (about $40 million) is dedicated to risk related research and development, the Woodrow Wilson Centre estimate that only around $11 million is actually directed towards risk related research. They argued in 2007 that it would be necessary to increase funding to a minimum of $50 million in the following two years so as to fill the gaps in knowledge in these areas.

The potential for workplace exposure was highlighted by the 2004 Royal Society report which recommended a review of existing regulations to assess and control workplace exposure to nanoparticles and nanotubes. The report expressed particular concern for the inhalation of large quantities of nanoparticles by workers involved in the manufacturing process.

Stakeholders concerned by the lack of a regulatory framework to assess and control risks associated with the release of nanoparticles and nanotubes have drawn parallels with bovine spongiform encephalopathy ('mad cow's disease'), thalidomide, genetically modified food, nuclear energy, reproductive technologies, biotechnology, and asbestosis. In light of such concerns, the Canadian based ETC Group have called for a moratorium on nano-related research until comprehensive regulatory frameworks are developed that will ensure workplace safety.

## Reactive Oxygen Species

For some types of particles, the smaller they are, the greater their surface area to volume ratio and the higher their chemical reactivity and biological activity. The greater chemical reactivity of nanomaterials can result in increased production of reactive oxygen species (ROS), including free radicals. ROS production has been found in a diverse range of nanomaterials including carbon fullerenes, carbon nanotubes and nanoparticle metal oxides. ROS and free radical production is one of the primary mechanisms of nanoparticle toxicity; it may result in oxidative stress, inflammation, and consequent damage to proteins, membranes and DNA.

## Biodistribution

The extremely small size of nanomaterials also means that they much more readily gain entry into the human body than larger sized particles. How these nanoparticles behave inside the body is still a major question that needs to be resolved. The behavior of nanoparticles is a function of their size, shape and surface reactivity with the surrounding tissue. In principle, a large number of particles could overload the body's phagocytes, cells that ingest and destroy foreign matter, thereby triggering stress reactions that lead to inflammation and weaken the body's defense against other pathogens. In addition to questions about what happens if non-degradable or slowly degradable nanoparticles accumulate in bodily organs, another concern is their potential interaction or interference with biological processes inside the body. Because of their large surface area, nanoparticles will, on exposure to tissue and fluids, immediately adsorb onto their surface some of the macromolecules they encounter. This may, for instance, affect the regulatory mechanisms of enzymes and other proteins.

Nanomaterials are able to cross biological membranes and access cells, tissues and organs that larger-sized particles normally cannot. Nanomaterials can gain access to the blood stream via inhalation or ingestion. At least some nanomaterials can penetrate the skin; even larger microparticles may penetrate skin when it is flexed. Broken skin is an ineffective particle barrier, suggesting that acne, eczema, shaving wounds or severe sunburn may accelerate skin uptake of nanomaterials. Then, once in the blood stream, nanomaterials can be transported around the body and be taken up by organs and tissues, including the brain, heart, liver, kidneys, spleen, bone marrow and nervous system. Nanomaterials have proved toxic to human tissue and cell cultures, resulting in increased oxidative stress, inflammatory cytokine production and cell death. Unlike larger particles, nanomaterials may be taken up by cell mitochondria and the cell nucleus. Studies demonstrate the potential for nanomaterials to cause DNA mutation and induce major structural damage to mitochondria, even resulting in cell death.

## Nanotoxicity Studies

There is presently no authority to specifically regulate nanotech-based products. Scientific research has indicated the potential for some nanomaterials to be toxic to humans or the environment. In March 2004 tests conducted by environmental toxicologist Eva Oberdörster, Ph.D. working with Southern Methodist University in Texas, found extensive brain damage to fish exposed to fullerenes for a period of just 48 hours at a relatively moderate dose of 0.5 parts per million (commensurate with levels of other kinds of pollution found in bays). The fish also exhibited changed gene markers in their livers, indicating their entire physiology was affected. In a concurrent test, the fullerenes killed water fleas, an important link in the marine food chain. The extremely small size of fabricated nanomaterials also means that they are much more readily taken up by living tissue than presently known toxins. Nanoparticles can be inhaled, swallowed, absorbed through skin and deliberately or accidentally injected during medical procedures. They might be accidentally or inadvertently released from materials implanted into living tissue.

Researcher Shosaku Kashiwada of the National Institute for Environmental Studies in Tsukuba, Japan, in a more recent study, intended to further investigate the effects of nanoparticles on soft-bodied organisms. His study allowed him to explore the distribution of water-suspended fluorescent nanoparticles throughout the eggs and adult bodies of a species of fish, known as the see-through medaka (Oryzias latipes). See-through medaka were used because of their small size, wide temperature and salinity tolerances, and short generation time. Moreover, small fish like the see-through medaka have been popular test subjects for human diseases and organogenesis for other reasons as well, including their transparent embryos, rapid embryo development, and the functional equivalence of their organs and tissue material to that of mammals. Because the see-through medaka have transparent bodies, analyzing the deposition of fluorescent nanoparticles throughout the body is quite simple. For his study, Dr. Kashiwada evaluated four aspects of nanoparticle accumulation. These included the overall accumulation and the size-dependent accumulation of nanoparticles by medaka eggs, the effects of salinity on the aggregation of nanoparticles in solution and on their accumulation by medaka eggs, and the distribution of nanoparticles in the blood and organs of adult medaka. It was also noted that nanoparticles were in fact taken up into the bloodstream and deposited throughout the body. In the medaka eggs, there was a high accumulation of nanoparticles in the yolk; most often bioavailibility was dependent on specific sizes of the particles. Adult samples of medaka had accumulated nanoparticles in the gills, intestine, brain, testis, liver, and bloodstream. One major result from this study was the fact that salinity may have a large influence on the bioavailibility and toxicity of nanoparticles to penetrate membranes and eventually kill the specimen.

As the use of nanomaterials increases worldwide, concerns for worker and user safety are mounting. To address such concerns, the Swedish Karolinska Institute conducted a study in which various nanoparticles were introduced to human lung epithelial cells. The results, released in 2008, showed that iron oxide nanoparticles caused little DNA damage and were non-toxic. Zinc oxide nanoparticles were slightly worse. Titanium dioxide caused only DNA damage. Carbon nanotubes caused DNA damage at low levels. Copper oxide was found to be the worst offender, and was the only nanomaterial identified by the researchers as a clear health risk. The latest toxicology studies on mice involving exposure to carbon nanotubes (CNT) showed a limited pulmonary inflammatory potential of MWCNT at levels corresponding to the average inhalable elemental carbon concentrations observed in U.S.-based CNT facilities. The study estimated that considerable years of exposure are necessary for significant pathology to occur.

## No Fullerene Toxicity Reported

Nanoparticles can also be made of $C_{60}$, as is the case with almost any room temperature solid, and several groups have done this and studied toxicity of such particles. The results in the work of Oberdörster at Southern Methodist University, published in "Environmental Health Perspectives" in July 2004, in which questions were raised of potential cytotoxicity, has now been shown by several sources to be likely caused by the tetrahydrofuran used in preparing the 30 nm–100 nm particles of $C_{60}$ used in the research. Isakovic, et al., 2006, who review this phenomenon, gives results showing that removal of THF from the $C_{60}$ particles resulted in a loss of toxicity. Sayes, et al., 2007, also show that particles prepared as in Oberdorster caused no detectable inflammatory response when instilled intratracheally in rats after observation for 3 months, suggesting that even the particles prepared by Oberdorster do not exhibit markers of toxicity in mammalian models. This work used as a benchmark quartz particles, which did give an inflammatory response.

A comprehensive and recent review of work on fullerene toxicity is available in "Toxicity Studies of Fullerenes and Derivatives," a chapter from the book "Bio-applications of Nanoparticles". In this work, the authors review the work on fullerene toxicity beginning in the early 1990s to present, and conclude that the evidence gathered since the discovery of fullerenes overwhelmingly points to $C_{60}$ being non-toxic. As is the case for toxicity profile with any chemical modification of a structural moiety, the authors suggest that individual molecules be assessed individually.

## Toxicity of Metal Based Nanoparticles

Metal based nanoparticles (NPs) are a prominent class of NPs synthesized for their functions as semiconductors, electroluminescents, and thermoelectric materials. Biomedically, these antibacterial NPs have been utilized in drug delivery systems to access areas previously inaccessible to conventional medicine. With the recent increase in interest and development of nanotechnology, many studies have been performed to assess whether the unique characteristics of these NPs, namely their small surface area to volume ratio, might negatively impact the environment upon which they were introduced. Researchers have since found that many metal and metal oxide NPs have detrimental effects on the cells with which they come into contact including but not limited to DNA breakage and oxidation, mutations, reduced cell viability, warped morphology, induced apoptosis and necrosis, and decreased proliferation.

## Cytotoxicity

A primary marker for the damaging effects of NPs has been cell viability as determined by state and exposed surface area of the cell membrane. Cells exposed to metallic NPs have, in the case of copper oxide, had up to 60% of their cells rendered unviable. When diluted, the positively charged metal ions often experience an electrostatic attraction to the cell membrane of nearby cells, covering the membrane and preventing it from permeating the necessary fuels and wastes. With less exposed membrane for transportation and communication, the cells are often rendered inactive.

NPs have been found to induce apoptosis in certain cells primarily due to the mitochondrial damage and oxidative stress brought on by the foreign NPs electrostatic reactions.

## Genotoxicity

Many methods ranging from comet assay to the HPRT gene mutation test have found that metal based NPs disrupt DNA and its replication process in a variety of cells. In a study examining the effects of nanosilver on DNA, AgNPs were introduced to lymphocyte cell DNA which was then examined for abnormalities. The exposure of the NPs correlated to a significant increase in micronuclei indicative of genetic fragmentation. Metal Oxides such as copper oxide, uraninite, and cobalt oxide have also been found to exert significant stress on exposed DNA. The damage done to the DNA will often result in mutated cells and colonies as found with the HPRT gene test.

## Coatings and Charges

NPs, in their implementation, are covered with coatings and sometimes given positive or negative charges depending upon the intended function. Studies have found that these external factors affect the degree of toxicity of NPs. Positive charges are usually found to amplify and cause of cellular damage much more noticeably than negative charges do. In a study in which b- and c-polyethylenimine coated AgNPs were attached to strands of Lambda DNA, the cationic b-polyethylenimine coated AgNP was found to lower the melting point of the DNA 50 °C lower than its anionic counterpart.

## Immunogenicity of Nanoparticles

Very little attention has been directed towards the potential immunogenicity of nano-structures. Nanostructures can activate the immune system, inducing inflammation, immune responses, allergy, or even affect to the immune cells in a deleterious or bene-ficial way (immunosuppression in autoimmune diseases, improving immune respons-es in vaccines). More studies are needed in order to know the potential deleterious or beneficial effects of nanostructures in the immune system. In comparison to conven-tional pharmaceutical agents, nanostructures have very large sizes, and immune cells, especially phagocytic cells, recognize and try to destroy them.

## Complications with Nanotoxicity Studies

Size is therefore a key factor in determining the potential toxicity of a particle. How-ever it is not the only important factor. Other properties of nanomaterials that influ-ence toxicity include: chemical composition, shape, surface structure, surface charge, aggregation and solubility, and the presence or absence of functional groups of other chemicals. The large number of variables influencing toxicity means that it is difficult to generalise about health risks associated with exposure to nanomaterials – each new nanomaterial must be assessed individually and all material properties must be taken into account.

In addition, standarization of toxicology tests between laboratories are needed. Díaz, B. *et al.* from the University of Vigo (Spain) has shown (Small, 2008) that many dif-ferent cell lines should be studied in order to know if a nanostructure induces toxicity, and human cells can internalize aggregated nanoparticles. Moreover, it is important to take into account that many nanostructures aggregate in biological fluids, but groups manufacturing nanostructures do not care much about this matter. Many efforts of in-terdisciplinary groups are strongly needed in order to progress in this field.

## Effect of Aggregation or Agglomeration of Nanoparticles

Many nanoparticles agglomerate or aggregate when they are placed in environmental or biological fluids. The terms agglomeration and aggregation have distinct definitions according to the standards organizations ISO and ASTM, where agglomeration signi-fies more loosely bound particles and aggregation signifies very tightly bound or fused particles (typically occurring during synthesis or drying). Nanoparticles frequently ag-glomerate due to the high ionic strength of environmental and biological fluids, which shields the repulsion due to charges on the nanoparticles. Unfortunately, agglomera-tion has frequently been ignored in nanotoxicity studies, even though agglomeration would be expected to affect nanotoxicity since it changes the size, surface area, and sedimentation properties of the nanoparticles. In addition, many nanoparticles will agglomerate to some extent in the environment or in the body before they reach their target, so it is desirable to study how toxicity is affected by agglomeration.

A method was published that can be used to produce different mean sizes of stable agglomerates of several metal, metal oxide, and polymer nanoparticles in cell culture media for cell toxicity studies. Different mean sizes of agglomerates are produced by allowing the nanoparticles to agglomerate to a particular size in cell culture media without protein, and then adding protein to coat the agglomerates and "freeze" them at that size. By waiting different amounts of time before adding protein, different mean sizes of agglomerates of a single type of nanoparticle can be produced in an otherwise identical solution, allowing one to study how agglomerate size affects toxicity. In addition, it was found that vortexing while adding a high concentration of nanoparticles to the cell culture media produces much less agglomerated nanoparticles than if the dispersed solution is only mixed after adding the nanoparticles.

The agglomeration/deagglomeration (mechanical stability) potentials of airborne engineered nanoparticle clusters also have significant influences on their size distribution profiles at the end-point of their environmental transport routes. Different aerosolization and deagglomeration systems have been established to test stability of nanoparticle agglomerates. For example, laboratory setups based on critical orifices have been used to apply a wide range of external shear forces onto airborne nanoparticles. After applying shear forces, the particle mean size decreased while the particle generation rate increased. In another pioneering study, four powder aerosolization systems (dustiness testing systems) were compared for the first time for their characteristics linked to aerosol generation.

## Challenges of the Nano-visualisation and Related Unknowns in Nanotoxicology

With comparison to more conventional toxicology studies, the nanotoxicology field is however suffering from a lack of easy characterisation of the potential contaminants, the "nano" scale being a scale difficult to comprehend. The biological systems are themselves still not completely known at this scale. Ultimate Atomic visualisation methods such as Electron microscopy (SEM and TEM) and Atomic force microscopy (AFM) analysis allow visualisation of the nano world. Further nanotoxicology studies will require precise characterisation of the specificities of a given nano-element : size, chemical composition, detailed shape, level of aggregation, combination with other vectors, etc. Above all, these properties would have to be determined not only on the nanocomponent before its introduction in the living environment but also in the (mostly aqueous) biological environment.

There is a need for new methodologies to quickly assess the presence and reactivity of nanoparticles in commercial, environmental, and biological samples since current detection techniques require expensive and complex analytical instrumentation. There have been recent attempts to address these issues by developing and investigating sensitive, simple and portable colorimetric detection assays that assess for the surface reactivity of NPs, which can be used to detect the presence of NPs, in environmental and biological relevant samples. Surface redox reactivity is a key emerging property related to potential toxicity of NPs with living cells, and can be used as a key surrogate for determine for the presence of NPs and a first tier analytical strategy toward assessing NP exposures.

It is difficult to determine the degree of effect of a specific nanoparticle when compared to those of comparable nanoparticles already present in our natural environment .

AEM - Analytical Electron Microscopy was used over 40 years ago to investigate amphibole asbestos bodies in Lake Superior from the Reserve Mining operations. This could non-destructively characterise sub-micron particles. Today AEM can fully characterise to atom dimensions.

# Nanoelectronics

Nanoelectronics refer to the use of nanotechnology in electronic components. The term covers a diverse set of devices and materials, with the common characteristic that they are so small that inter-atomic interactions and quantum mechanical properties need to be studied extensively. Some of these candidates include: hybrid molecular/semiconductor electronics, one-dimensional nanotubes/nanowires, or advanced molecular electronics. Recent silicon CMOS technology generations, such as the 22 nanometer node, are already within this regime. Nanoelectronics are sometimes considered as disruptive technology because present candidates are significantly different from traditional transistors.

## Fundamental Concepts

In 1965 Gordon Moore observed that silicon transistors were undergoing a continual process of scaling downward, an observation which was later codified as Moore's law. Since his observation transistor minimum feature sizes have decreased from 10 micrometers to the 28-22 nm range in 2011. The field of nanoelectronics aims to enable the continued realization of this law by using new methods and materials to build electronic devices with feature sizes on the nanoscale.

The volume of an object decreases as the third power of its linear dimensions, but the surface area only decreases as its second power. This somewhat subtle and unavoidable principle has huge ramifications. For example, the power of a drill (or any other machine) is proportional to the volume, while the friction of the drill's bearings and gears is proportional to their surface area. For a normal-sized drill, the power of the device is enough to handily overcome any friction. However, scaling its length down by a factor of 1000, for example, decreases its power by $1000^3$ (a factor of a billion) while reducing the friction by only $1000^2$ (a factor of only a million). Proportionally it has 1000 times less power per unit friction than the original drill. If the original friction-to-power ratio was, say, 1%, that implies the smaller drill will have 10 times as much friction as power; the drill is useless.

For this reason, while super-miniature electronic integrated circuits are fully functional, the same technology cannot be used to make working mechanical devices beyond the scales where frictional forces start to exceed the available power. So even though you may see microphotographs of delicately etched silicon gears, such devices are currently little more than curiosities with limited real world applications, for example, in moving mirrors and shutters. Surface tension increases in much the same way, thus magnifying the tendency for very small objects to stick together. This could possibly make any kind of "micro factory" impractical: even if robotic arms and hands could be scaled down, anything they pick up will tend to be impossible to put down. The above being said, molecular evolution has resulted in working cilia, flagella, muscle fibers and rotary motors in aqueous environments, all on the nanoscale. These machines exploit the increased frictional forces found at the micro or nanoscale. Unlike a paddle or a propeller which depends on normal frictional forces (the frictional forces perpendicular to the surface) to achieve propulsion, cilia develop motion from the exaggerated drag or laminar forces (frictional forces parallel to the surface) present at micro and nano dimensions. To build meaningful "machines" at the nanoscale, the relevant forces need to be considered. We are faced with the development and design of intrinsically pertinent machines rather than the simple reproductions of macroscopic ones.

All scaling issues therefore need to be assessed thoroughly when evaluating nanotechnology for practical applications.

## Approaches to Nanoelectronics

## Nanofabrication

For example, single electron transistors, which involve transistor operation based on a single electron. Nanoelectromechanical systems also fall under this category. Nanofabrication can be used to construct ultradense parallel arrays of nanowires, as an alternative to synthesizing nanowires individually.

## Nanomaterials Electronics

Besides being small and allowing more transistors to be packed into a single chip, the uniform and symmetrical structure of nanotubes allows a higher electron mobility (faster electron movement in the material), a higher dielectric constant (faster frequency), and a symmetrical electron/hole characteristic.

Also, nanoparticles can be used as quantum dots.

## Molecular Electronics

Single molecule devices are another possibility. These schemes would make heavy use of molecular self-assembly, designing the device components to construct a larger structure or even a complete system on their own. This can be very useful for reconfigurable computing, and may even completely replace present FPGA technology.

Molecular electronics is a new technology which is still in its infancy, but also brings hope for truly atomic scale electronic systems in the future. One of the more promising applications of molecular electronics was proposed by the IBM researcher Ari Aviram and the theoretical chemist Mark Ratner in their 1974 and 1988 papers *Molecules for Memory, Logic and Amplification*.

This is one of many possible ways in which a molecular level diode / transistor might be synthesized by organic chemistry. A model system was proposed with a spiro carbon structure giving a molecular diode about half a nanometre across which could be connected by polythiophene molecular wires. Theoretical calculations showed the design to be sound in principle and there is still hope that such a system can be made to work.

## Other Approaches

Nanoionics studies the transport of ions rather than electrons in nanoscale systems.

Nanophotonics studies the behavior of light on the nanoscale, and has the goal of developing devices that take advantage of this behavior.

## Nanoelectronic Devices

Current high-technology production processes are based on traditional top down strategies, where nanotechnology has already been introduced silently. The critical length scale of integrated circuits is already at the nanoscale (50 nm and below) regarding the gate length of transistors in CPUs or DRAM devices.

## Computers

Nanoelectronics holds the promise of making computer processors more powerful than are possible with conventional semiconductor fabrication techniques. A number of approaches are currently being researched, including new forms of nanolithography, as well as the use of nanomaterials such as nanowires or small molecules in place of traditional CMOS components. Field effect transistors have been made using both semiconducting carbon nanotubes and with heterostructured semiconductor nanowires.

In 1999, the CMOS transistor developed at the Laboratory for Electronics and Information Technology in Grenoble, France, tested the limits of the principles of the MOSFET transistor with a diameter of 18 nm (approximately 70 atoms placed side by side). This was almost one tenth the size of the smallest industrial transistor in 2003 (130 nm in 2003, 90 nm in 2004, 65 nm in 2005 and 45 nm in 2007). It enabled the theoretical integration of seven billion junctions on a €1 coin. However, the CMOS transistor, which was created in 1999, was not a simple research experiment to study how CMOS technology functions, but rather a demonstration of how this technology functions now that we ourselves are getting ever closer to working on a molecular scale. Today it would be impossible to master the coordinated assembly of a large number of these transistors on a circuit and it would also be impossible to create this on an industrial level.

## Memory Storage

Electronic memory designs in the past have largely relied on the formation of transistors. However, research into crossbar switch based electronics have offered an alternative using reconfigurable interconnections between vertical and horizontal wiring arrays to create ultra high density memories. Two leaders in this area are Nantero which has developed a carbon nanotube based crossbar memory called Nano-RAM and Hewlett-Packard which has proposed the use of memristor material as a future replacement of Flash memory.

An example of such novel devices is based on spintronics. The dependence of the resistance of a material (due to the spin of the electrons) on an external field is called magnetoresistance. This effect can be significantly amplified (GMR - Giant Magneto-Resistance) for nanosized objects, for example when two ferromagnetic layers are separated by a nonmagnetic layer, which is several nanometers thick (e.g. Co-Cu-Co). The GMR effect has led to a strong increase in the data storage density of hard disks and made the gigabyte range possible. The so-called tunneling magnetoresistance (TMR) is very similar to GMR and based on the spin dependent tunneling of electrons through adjacent ferromagnetic layers. Both GMR and TMR effects can be used to create a non-volatile main memory for computers, such as the so-called magnetic random access memory or MRAM.

## Novel Optoelectronic Devices

In the modern communication technology traditional analog electrical devices are increasingly replaced by optical or optoelectronic devices due to their enormous bandwidth and capacity, respectively. Two promising examples are photonic crystals and quantum dots. Photonic crystals are materials with a periodic variation in the refractive index with a lattice constant that is half the wavelength of the light used. They offer a selectable band gap for the propagation of a certain wavelength, thus they resemble a semiconductor, but for light or photons instead of electrons. Quantum dots are nanoscaled objects, which can be used, among many other things, for the construction of lasers. The advantage of a quantum dot laser over the traditional semiconductor laser is that their emitted wavelength depends on the diameter of the dot. Quantum dot lasers are cheaper and offer a higher beam quality than conventional laser diodes.

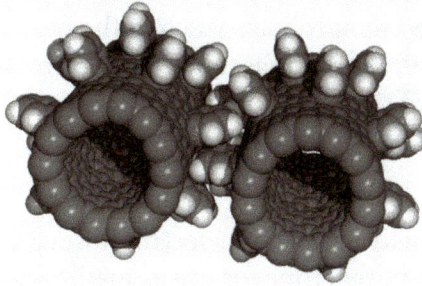

## Displays

The production of displays with low energy consumption might be accomplished using carbon nanotubes (CNT). Carbon nanotubes are electrically conductive and due to their small diameter of several nanometers, they can be used as field emitters with extremely high efficiency for field emission displays (FED). The principle of operation resembles that of the cathode ray tube, but on a much smaller length scale.

## Quantum Computers

Entirely new approaches for computing exploit the laws of quantum mechanics for novel quantum computers, which enable the use of fast quantum algorithms. The Quantum computer has quantum bit memory space termed "Qubit" for several computations at the same time. This facility may improve the performance of the older systems.

## Radios

Nanoradios have been developed structured around carbon nanotubes.

## Energy Production

Research is ongoing to use nanowires and other nanostructured materials with the hope to create cheaper and more efficient solar cells than are possible with conventional planar silicon solar cells. It is believed that the invention of more efficient solar energy would have a great effect on satisfying global energy needs.

There is also research into energy production for devices that would operate *in vivo*, called bio-nano generators. A bio-nano generator is a nanoscale electrochemical device, like a fuel cell or galvanic cell, but drawing power from blood glucose in a living body, much the same as how the body generates energy from food. To achieve the effect, an enzyme is used that is capable of stripping glucose of its electrons, freeing them for use in electrical devices. The average person's body could, theoretically, generate 100 watts of electricity (about 2000 food calories per day) using a bio-nano generator. However, this estimate is only true if all food was converted to electricity, and the human body needs some energy consistently, so possible power generated is likely much lower. The electricity generated by such a device could power devices embedded in the body (such as pacemakers), or sugar-fed nanorobots. Much of the research done on bio-nano generators is still experimental, with Panasonic's Nanotechnology Research Laboratory among those at the forefront.

## Medical Diagnostics

There is great interest in constructing nanoelectronic devices that could detect the concentrations of biomolecules in real time for use as medical diagnostics, thus falling into the category of nanomedicine. A parallel line of research seeks to create nanoelectronic devices which could interact with single cells for use in basic biological research. These devices are called nanosensors. Such miniaturization on nanoelectronics towards in vivo proteomic sensing should enable new approaches for health monitoring, surveillance, and defense technology.

# Green Nanotechnology

Green nanotechnology refers to the use of nanotechnology to enhance the environmental sustainability of processes producing negative externalities. It also refers to the use of the products of nanotechnology to enhance sustainability. It includes making green nano-products and using nano-products in support of sustainability.

Green nanotechnology has been described as the development of clean technologies, "to minimize potential environmental and human health risks associated with the manufacture and use of nanotechnology products, and to encourage replacement of existing products with new nano-products that are more environmentally friendly throughout their lifecycle."

## Goals

Green nanotechnology has two goals: producing nanomaterials and products without harming the environment or human health, and producing nano-products that provide solutions to environmental problems. It uses existing principles of green chemistry and green engineering to make nanomaterials and nano-products without toxic ingredients, at low temperatures using less energy and renewable inputs wherever possible, and using lifecycle thinking in all design and engineering stages.

In addition to making nanomaterials and products with less impact to the environment, green nanotechnology also means using nanotechnology to make current manufacturing processes for non-nano materials and products more environmentally friendly. For example, nanoscale membranes can help separate desired chemical reaction products from waste materials. Nanoscale catalysts can make chemical reactions more efficient and less wasteful. Sensors at the nanoscale can form a part of process control systems, working with nano-enabled information systems. Using alternative energy systems, made possible by nanotechnology, is another way to "green" manufacturing processes.

The second goal of green nanotechnology involves developing products that benefit the environment either directly or indirectly. Nanomaterials or products directly can clean hazardous waste sites, desalinate water, treat pollutants, or sense and monitor environmental pollutants. Indirectly, lightweight nanocomposites for automobiles and other means of transportation could save fuel and reduce materials used for production; nanotechnology-enabled fuel cells and light-emitting diodes (LEDs) could reduce pollution from energy generation and help conserve fossil fuels; self-cleaning nanoscale surface coatings could reduce or eliminate many cleaning chemicals used in regular maintenance routines; and enhanced battery life could lead to less material use and less waste. Green Nanotechnology takes a broad systems view of nanomaterials and products, ensuring that unforeseen consequences are minimized and that impacts are anticipated throughout the full life cycle.

## Current Research

## Solar Cells

Research is underway to use nanomaterials for purposes including more efficient solar cells, practical fuel cells, and environmentally friendly batteries. The most advanced nanotechnology projects related to energy are: storage, conversion, manufacturing improvements by reducing materials and process rates, energy saving (by better thermal insulation for example), and enhanced renewable energy sources.

One major project that is being worked on is the development of nanotechnology in solar cells. Solar cells are more efficient as they get tinier and solar energy is a renewable resource. The price per watt of solar energy is lower than one dollar.

Research is ongoing to use nanowires and other nanostructured materials with the hope of to create cheaper and more efficient solar cells than are possible with conventional planar silicon solar cells. Another example is the use of fuel cells powered by hydrogen, potentially using a catalyst consisting of carbon supported noble metal particles with diameters of 1-5 nm. Materials with small nanosized pores may be suitable for hydrogen storage. Nanotechnology may also find applications in batteries, where the use of nanomaterials may enable batteries with higher energy content or supercapacitors with a higher rate of recharging.

Nanotechnology is already used to provide improved performance coatings for photovoltaic (PV) and solar thermal panels. Hydrophobic and self-cleaning properties combine to create more efficient solar panels, especially during inclement weather. PV covered with nanotechnology coatings are said to stay cleaner for longer to ensure maximum energy efficiency is maintained.

## Nanoremediation and Water Treatment

Nanotechnology offers the potential of novel nanomaterials for the treatment of surface water, groundwater, wastewater, and other environmental materials contaminated by toxic metal ions, organic and inorganic solutes, and microorganisms. Due to their unique activity toward recalcitrant contaminants, many nanomaterials are under active research and development for use in the treatment of water and contaminated sites.

The present market of nanotech-based technologies applied in water treatment consists of reverse osmosis, nanofiltration, ultrafiltration membranes. Indeed, among emerging products one can name nanofiber filters, carbon nanotubes and various nanoparticles. Nanotechnology is expected to deal more efficiently with contaminants which convectional water treatment systems struggle to treat, including bacteria, viruses and heavy metals. This efficiency generally stems from the very high specific surface area of nanomaterials which increases dissolution, reactivity and sorption of contaminants.

Some potential applications include:

- To maintain public health, pathogens in water need to be identified rapidly and reliably. Unfortunately, traditional laboratory culture tests take days to complete. Faster methods involving enzymes, immunological or genetic tests are under development.

- Water filtration may be improved with the use of nanofiber membranes and the use of nanobiocides, which appear promisingly effective.

- Biofilms are mats of bacteria wrapped in natural polymers. These can be difficult to treat with antimicrobials or other chemicals. They can be cleaned up mechanically, but at the cost of substantial down-time and labour. Work is in progress to develop enzyme treatments that may be able to break down such biofilms.

## Environmental Remediation

Nanoremediation is the use of nanoparticles for environmental remediation. Nanoremediation has been most widely used for groundwater treatment, with additional extensive research in wastewater treatment. Nanoremediation has also been tested for soil and sediment cleanup. Even more preliminary research is exploring the use of nanoparticles to remove toxic materials from gases.

Some nanoremediation methods, particularly the use of nano zero-valent iron for groundwater cleanup, have been deployed at full-scale cleanup sites. Nanoremediation is an emerging industry; by 2009, nanoremediation technologies had been documented in at least 44 cleanup sites around the world, predominantly in the United States. During nanoremediation, a nanoparticle agent must be brought into contact with the target contaminant under conditions that allow a detoxifying or immobilizing reaction. This process typically involves a pump-and-treat process or *in situ* application. Other methods remain in research phases.

Scientists have been researching the capabilities of buckminsterfullerene in controlling pollution, as it may be able to control certain chemical reactions. Buckminsterfullerene has been demonstrated as having the ability of inducing the protection of reactive oxygen species and causing lipid peroxidation. This material may allow for hydrogen fuel to be more accessible to consumers.

## Water Filtration

Nanofiltration is a relatively recent membrane filtration process used most often with low total dissolved solids water such as surface water and fresh groundwater, with the purpose of softening (polyvalent cation removal) and removal of disinfection by-product precursors such as natural organic matter and synthetic organic matter. Nanofiltration is also becoming more widely used in food processing applications such as dairy, for simultaneous concentration and partial (monovalent ion) demineralisation.

Nanofiltration is a membrane filtration based method that uses nanometer sized cylindrical through-pores that pass through the membrane at a 90°. Nanofiltration membranes have pore sizes from 1-10 Angstrom, smaller than that used in microfiltration and ultrafiltration, but just larger than that in reverse osmosis. Membranes used are predominantly created from polymer thin films. Materials that are commonly used include polyethylene terephthalate or metals such as aluminum. Pore dimensions are controlled by pH, temperature and time during development with pore densities ranging from 1 to 106 pores per cm$^2$. Membranes made from polyethylene terephthalate and other similar materials, are referred to as "track-etch" membranes, named after the way the pores on the membranes are made. "Tracking" involves bombarding the polymer thin film with high energy particles. This results in making tracks that are chemically developed into the membrane, or "etched" into the membrane, which are the pores. Membranes created from metal such as alumina membranes, are made by electrochemically growing a thin layer of aluminum oxide from aluminum metal in an acidic medium.

Some water-treatment devices incorporating nanotechnology are already on the market, with more in development. Low-cost nanostructured separation membranes methods have been shown to be effective in producing potable water in a recent study.

# DNA Nanotechnology

DNA nanotechnology is the design and manufacture of artificial nucleic acid structures for technological uses. In this field, nucleic acids are used as non-biological engineering materials for nanotechnology rather than as the carriers of genetic information in living cells. Researchers in the field have created static structures such as two- and three-dimensional crystal lattices, nanotubes, polyhedra, and arbitrary shapes, and functional devices such as molecular machines and DNA computers. The field is beginning to be used as a tool to solve basic science problems in structural biology and biophysics, including applications in X-ray crystallography and nuclear magnetic resonance spectroscopy of proteins to determine structures. Potential applications in molecular scale electronics and nanomedicine are also being investigated.

The conceptual foundation for DNA nanotechnology was first laid out by Nadrian Seeman in the early 1980s, and the field began to attract widespread interest in the mid-2000s. This use of nucleic acids is enabled by their strict base pairing rules, which cause only portions of strands with complementary base sequences to bind together to form strong, rigid double helix structures. This allows for the rational design of base sequences that will selectively assemble to form complex target structures with precisely controlled nanoscale features. Several assembly methods are used to make these structures, including tile-based structures that assemble from smaller structures, folding structures using the DNA origami method, and dynamically reconfigurable structures using strand displacement methods. While the field's name specifically references DNA, the same principles have been used with other types of nucleic acids as well, leading to the occasional use of the alternative name *nucleic acid nanotechnology*.

## Fundamental Concepts

These four strands associate into a DNA four-arm junction because this structure maximizes the number of correct base pairs, with A matched to T and C matched to G. See this image for a more realistic model of the four-arm junction showing its tertiary structure.

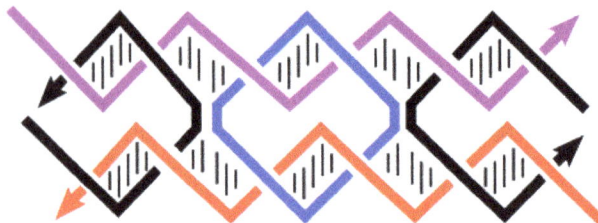

This double-crossover (DX) supramolecular complex consists of five DNA single strands that form two double-helical domains, on the top and the bottom in this image. There are two crossover points where the strands cross from one domain into the other.

## Properties of Nucleic Acids

Nanotechnology is often defined as the study of materials and devices with features on a scale below 100 nanometers. DNA nanotechnology, specifically, is an example of bottom-up molecular self-assembly, in which molecular components spontaneously organize into stable structures; the particular form of these structures is induced by the physical and chemical properties of the components selected by the designers. In DNA nanotechnology, the component materials are strands of nucleic acids such as DNA; these strands are often synthetic and are almost always used outside the context of a living cell. DNA is well-suited to nanoscale construction because the binding between two nucleic acid strands depends on simple base pairing rules which are well understood, and form the specific nanoscale structure of the nucleic acid double helix. These qualities make the assembly of nucleic acid structures easy to control through nucleic acid design. This property is absent in other materials used in nanotechnology, including proteins, for which protein design is very difficult, and nanoparticles, which lack the capability for specific assembly on their own.

The structure of a nucleic acid molecule consists of a sequence of nucleotides distinguished by which nucleobase they contain. In DNA, the four bases present are adenine (A), cytosine (C), guanine (G), and thymine (T). Nucleic acids have the property that two molecules will only bind to each other to form a double helix if the two sequences are complementary, meaning that they form matching sequences of base pairs, with A only binding to T, and C only to G. Because the formation of correctly matched base pairs is energetically favorable, nucleic acid strands are expected in most cases to bind to each other in the conformation that maximizes the number of correctly paired bases. The sequences of bases in a system of strands thus determine the pattern of binding and the overall structure in an easily controllable way. In DNA nanotechnology, the base sequences of strands are rationally designed by researchers so that the base pairing interactions cause the strands to assemble in the desired conformation. While DNA is the dominant material used, structures incorporating other nucleic acids such as RNA and peptide nucleic acid (PNA) have also been constructed.

## Subfields

DNA nanotechnology is sometimes divided into two overlapping subfields: structural DNA nanotechnology and dynamic DNA nanotechnology. Structural DNA nanotechnology, sometimes abbreviated as SDN, focuses on synthesizing and characterizing nucleic acid complexes and materials that assemble into a static, equilibrium end state. On the other hand, dynamic DNA nanotechnology focuses on complexes with useful non-equilibrium behavior such as the ability to reconfigure based on a chemical or physical stimulus. Some complexes, such as nucleic acid nanomechanical devices, combine features of both the structural and dynamic subfields.

The complexes constructed in structural DNA nanotechnology use topologically branched nucleic acid structures containing junctions. (In contrast, most biological DNA exists as an unbranched double helix.) One of the simplest branched structures is a four-arm junction that consists of four individual DNA strands, portions of which are complementary in a specific pattern. Unlike in natural Holliday junctions, each arm in the artificial immobile four-arm junction has a different base sequence, causing the junction point to be fixed at a certain position. Multiple junctions can be combined in the same complex, such as in the widely used double-crossover (DX) structural motif, which contains two parallel double helical domains with individual strands crossing between the domains at two crossover points. Each crossover point is, topologically, a four-arm junction, but is constrained to one orientation, in contrast to the flexible single four-arm junction, providing a rigidity that makes the DX motif suitable as a structural building block for larger DNA complexes.

Dynamic DNA nanotechnology uses a mechanism called toehold-mediated strand displacement to allow the nucleic acid complexes to reconfigure in response to the addition of a new nucleic acid strand. In this reaction, the incoming strand binds to a single-stranded toehold region of a double-stranded complex, and then displaces one of the strands bound in the original complex through a branch migration process. The overall effect is that one of the strands in the complex is replaced with another one. In addition, reconfigurable structures and devices can be made using functional nucleic acids such as deoxyribozymes and ribozymes, which can perform chemical reactions, and aptamers, which can bind to specific proteins or small molecules.

## Structural DNA Nanotechnology

Structural DNA nanotechnology, sometimes abbreviated as SDN, focuses on synthesizing and characterizing nucleic acid complexes and materials where the assembly has a static, equilibrium endpoint. The nucleic acid double helix has a robust, defined three-dimensional geometry that makes it possible to predict and design the structures of more complicated nucleic acid complexes. Many such structures have been created, including two- and three-dimensional structures, and periodic, aperiodic, and discrete structures.

## Extended Lattices

The assembly of a DX array. *Left*, schematic diagram. Each bar represents a double-helical domain of DNA, with the shapes representing complementary sticky ends. The DX complex at top will combine with other DX complexes into the two-dimensional array shown at bottom. *Right*, an atomic force microscopy image of the assembled array. The individual DX tiles are clearly visible within the assembled structure. The field is 150 nm across.

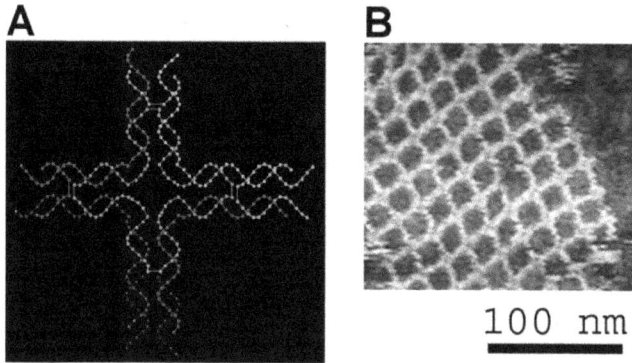

*Left*, a model of a DNA tile used to make another two-dimensional periodic lattice. *Right*, an atomic force micrograph of the assembled lattice.

An example of an aperiodic two-dimensional lattice that assembles into a fractal pattern. *Left*, the Sierpinski gasket fractal. *Right*, DNA arrays that display a representation of the Sierpinski gasket on their surfaces

Small nucleic acid complexes can be equipped with sticky ends and combined into larger two-dimensional periodic lattices containing a specific tessellated pattern of the individual molecular tiles. The earliest example of this used double-crossover (DX) complexes as the basic tiles, each containing four sticky ends designed with sequences that caused the DX units to combine into periodic two-dimensional flat sheets that are essentially rigid two-dimensional crystals of DNA. Two-dimensional arrays have been made from other motifs as well, including the Holliday junction rhombus lattice, and various DX-based arrays making use of a double-cohesion scheme. The top two images at right show examples of tile-based periodic lattices.

Two-dimensional arrays can be made to exhibit aperiodic structures whose assembly implements a specific algorithm, exhibiting one form of DNA computing. The DX tiles can have their sticky end sequences chosen so that they act as Wang tiles, allowing them to perform computation. A DX array whose assembly encodes an XOR operation has been demonstrated; this allows the DNA array to implement a cellular automaton that generates a fractal known as the Sierpinski gasket. The third image at right shows this type of array. Another system has the function of a binary counter, displaying a representation of increasing binary numbers as it grows. These results show that computation can be incorporated into the assembly of DNA arrays.

DX arrays have been made to form hollow nanotubes 4–20 nm in diameter, essentially two-dimensional lattices which curve back upon themselves. These DNA nanotubes are somewhat similar in size and shape to carbon nanotubes, and while they lack the electrical conductance of carbon nanotubes, DNA nanotubes are more easily modified and connected to other structures. One of many schemes for constructing DNA nanotubes uses a lattice of curved DX tiles that curls around itself and closes into a tube. In an alternative method that allows the circumference to be specified in a simple, modular fashion using single-stranded tiles, the rigidity of the tube is an emergent property.

Forming three-dimensional lattices of DNA was the earliest goal of DNA nanotechnology, but this proved to be one of the most difficult to realize. Success using a motif based on the concept of tensegrity, a balance between tension and compression forces, was finally reported in 2009.

## Discrete Structures

Researchers have synthesized many three-dimensional DNA complexes that each have the connectivity of a polyhedron, such as a cube or octahedron, meaning that the DNA duplexes trace the edges of a polyhedron with a DNA junction at each vertex. The earliest demonstrations of DNA polyhedra were very work-intensive, requiring multiple ligations and solid-phase synthesis steps to create catenated polyhedra. Subsequent work yielded polyhedra whose synthesis was much easier. These include a DNA octahedron made from a long single strand designed to fold into the correct conformation, and a tetrahedron that can be produced from four DNA strands in one step, pictured at the top of this article.

Nanostructures of arbitrary, non-regular shapes are usually made using the DNA origami method. These structures consist of a long, natural virus strand as a "scaffold", which is made to fold into the desired shape by computationally designed short "staple" strands. This method has the advantages of being easy to design, as the base sequence is predetermined by the scaffold strand sequence, and not requiring high strand purity and accurate stoichiometry, as most other DNA nanotechnology methods do. DNA origami was first demonstrated for two-dimensional shapes, such as a smiley face and a coarse map of the Western Hemisphere. Solid three-dimensional structures can be made by using parallel DNA helices arranged in a honeycomb pattern, and structures with two-dimensional faces can be made to fold into a hollow overall three-dimensional shape, akin to a cardboard box. These can be programmed to open and reveal or release a molecular cargo in response to a stimulus, making them potentially useful as programmable molecular cages.

## Templated Assembly

Nucleic acid structures can be made to incorporate molecules other than nucleic acids, sometimes called heteroelements, including proteins, metallic nanoparticles, quantum dots, and fullerenes. This allows the construction of materials and devices with a range of functionalities much greater than is possible with nucleic acids alone. The goal is to use the self-assembly of the nucleic acid structures to template the assembly of the nanoparticles hosted on them, controlling their position and in some cases orientation. Many of these schemes use a covalent attachment scheme, using oligonucleotides with amide or thiol functional groups as a chemical handle to bind the heteroelements. This covalent binding scheme has been used to arrange gold nanoparticles on a DX-based array, and to arrange streptavidin protein molecules into specific patterns on a DX array. A non-covalent hosting scheme using Dervan polyamides on a DX array was used to arrange streptavidin proteins in a specific pattern on a DX array. Carbon nanotubes have been hosted on DNA arrays in a pattern allowing the assembly to act as a molecular electronic device, a carbon nanotube field-effect transistor. In addition, there are nucleic acid metallization methods, in which the nucleic acid is replaced by a metal which assumes the general shape of the original nucleic acid structure, and schemes for using nucleic acid nanostructures as lithography masks, transferring their pattern into a solid surface.

## Dynamic DNA Nanotechnology

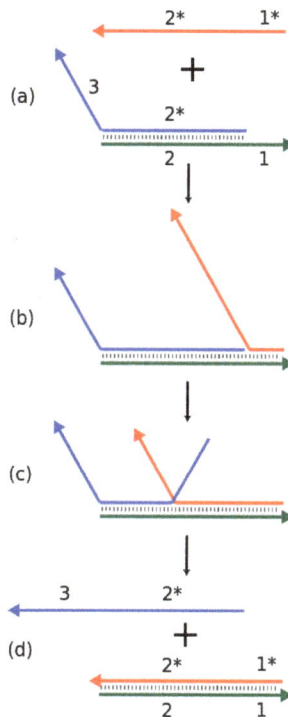

Dynamic DNA nanotechnology often makes use of toehold-mediated strand displacement reactions. In this example, the red strand binds to the single stranded toehold region on the green strand (region 1), and then in a branch migration process across region 2, the blue strand is displaced and freed from the complex. Reactions like these are used to dynamically reconfigure or assemble nucleic acid nanostructures. In addition, the red and blue strands can be used as signals in a molecular logic gate.

Dynamic DNA nanotechnology focuses on forming nucleic acid systems with designed dynamic functionalities related to their overall structures, such as computation and mechanical motion. There is some overlap between structural and dynamic DNA nanotechnology, as structures can be formed through annealing and then reconfigured dynamically, or can be made to form dynamically in the first place.

## Nanomechanical Devices

DNA complexes have been made that change their conformation upon some stimulus, making them one form of nanorobotics. These structures are initially formed in the same way as the static structures made in structural DNA nanotechnology, but are designed so that dynamic reconfiguration is possible after the initial assembly. The earliest such device made use of the transition between the B-DNA and Z-DNA forms to respond to a change in buffer conditions by undergoing a twisting motion. This reliance on buffer conditions, however, caused all devices to change state at the same time. Subsequent systems could change states based upon the presence of control strands, allowing multiple devices to be independently operated in solution. Some examples of such systems are a "molecular tweezers" design that has an open and a closed state, a device that could switch from a paranemic-crossover (PX) conformation to a double-junction (JX2) conformation, undergoing rotational motion in the process, and a two-dimensional array that could dynamically expand and contract in response to control strands. Structures have also been made that dynamically open or close, potentially acting as a molecular cage to release or reveal a functional cargo upon opening.

DNA walkers are a class of nucleic acid nanomachines that exhibit directional motion along a linear track. A large number of schemes have been demonstrated. One strategy is to control the motion of the walker along the track using control strands that need to be manually added in sequence. Another approach is to make use of restriction enzymes or deoxyribozymes to cleave the strands and cause the walker to move forward, which has the advantage of running autonomously. A later system could walk upon a two-dimensional surface rather than a linear track, and demonstrated the ability to selectively pick up and move molecular cargo. Additionally, a linear walker has been demonstrated that performs DNA-templated synthesis as the walker advances along the track, allowing autonomous multistep chemical synthesis directed by the walker. The synthetic DNA walkers' function is similar to that of the proteins dynein and kinesin.

## Strand Displacement Cascades

Cascades of strand displacement reactions can be used for either computational or structural purposes. An individual strand displacement reaction involves revealing a new sequence in response to the presence of some initiator strand. Many such reactions can be linked into a cascade where the newly revealed output sequence of one reaction can initiate another strand displacement reaction elsewhere. This in turn allows for the construction of chemical reaction networks with many components, exhibiting complex computational and information processing abilities. These cascades are made energetically favorable through the formation of new base pairs, and the entropy gain from disassembly reactions. Strand displacement cascades allow isothermal operation of the assembly or computational process, in contrast to traditional nucleic acid assembly's requirement for a thermal annealing step, where the temperature is raised and then slowly lowered to ensure proper formation of the desired structure. They can also support catalytic function of the initiator species, where less than one equivalent of the initiator can cause the reaction to go to completion.

Strand displacement complexes can be used to make molecular logic gates capable of complex computation. Unlike traditional electronic computers, which use electric current as inputs and outputs, molecular computers use the concentrations of specific chemical species as signals. In the case of nucleic acid strand displacement circuits, the signal is the presence of nucleic acid strands that are released or consumed by binding and unbinding events to other strands in displacement complexes. This approach has been used to make logic gates such as AND, OR, and NOT gates. More recently, a four-bit circuit was demonstrated that can compute the square root of the integers 0–15, using a system of gates containing 130 DNA strands.

Another use of strand displacement cascades is to make dynamically assembled structures. These use a hairpin structure for the reactants, so that when the input strand binds, the newly revealed sequence is on the same molecule rather than disassembling. This allows new opened hairpins to be added to a growing complex. This approach has been used to make simple structures such as three- and four-arm junctions and dendrimers.

## Applications

DNA nanotechnology provides one of the few ways to form designed, complex structures with precise control over nanoscale features. The field is beginning to see application to solve basic science problems in structural biology and biophysics. The earliest such application envisaged for the field, and one still in development, is in crystallography, where molecules that are difficult to crystallize in isolation could be arranged within a three-dimensional nucleic acid lattice, allowing determination of their structure. Another application is the use of DNA origami rods to replace liquid crystals in residual dipolar coupling experiments in protein NMR spectroscopy; using DNA origami is advantageous because, unlike liquid crystals, they are tolerant of the detergents needed to suspend membrane proteins in solution. DNA walkers have been used as nanoscale assembly lines to move nanoparticles and direct chemical synthesis. Further, DNA origami structures have aided in the biophysical studies of enzyme function and protein folding.

DNA nanotechnology is moving toward potential real-world applications. The ability of nucleic acid arrays to arrange other molecules indicates its potential applications in molecular scale electronics. The assembly of a nucleic acid structure could be used to template the assembly of a molecular electronic elements such as molecular wires, providing a method for nanometer-scale control of the placement and overall architecture of the device analogous to a molecular breadboard. DNA nanotechnology has been compared to the concept of programmable matter because of the coupling of computation to its material properties.

In a study conducted by a group of scientists from iNANO center and CDNA Center in Aarhus university (Aarhus), researchers were able to construct a small multi-switchable 3D DNA Box Origami. The proposed nanoparticle was characterized by atomic force microscopy (AFM), transmission electron microscopy (TEM) and Förster resonance energy transfer (FRET). The constructed box was shown to have a unique reclosing mechanism, which enabled it to repeatedly open and close in response to a unique set of DNA or RNA keys. The authors proposed that this "DNA device can potentially be used for a broad range of applications such as controlling the function of single molecules, controlled drug delivery, and molecular computing.".

There are potential applications for DNA nanotechnology in nanomedicine, making use of its ability to perform computation in a biocompatible format to make "smart drugs" for targeted drug delivery. One such system being investigated uses a hollow DNA box containing proteins that induce apoptosis, or cell death, that will only open when in proximity to a cancer cell. There has additionally been interest in expressing these artificial structures in engineered living bacterial cells, most likely using the transcribed RNA for the assembly, although it is unknown whether these complex structures are able to efficiently fold or assemble in the cell's cytoplasm. If successful, this could enable directed evolution of nucleic acid nanostructures. Scientists at Oxford University reported the self-assembly of four short strands of synthetic DNA into a cage which can enter cells and survive for at least 48 hours. The fluorescently labeled DNA tetrahedra were found to remain intact in the laboratory cultured human kidney cells despite the attack by cellular enzymes after two days. This experiment showed the potential of drug delivery inside the living cells using the DNA 'cage'. A DNA tetrahedron was used to deliver RNA Interference (RNAi) in a mouse model, reported a team of researchers in MIT. Delivery of the interfering RNA for treatment has showed some success using polymer or lipid, but there are limits of safety and imprecise targeting, in addition to short shelf life in the blood stream. The DNA nanostructure created by the team consists of six strands of DNA to form a tetrahedron, with one strand of RNA affixed to each of the six edges. The tetrahedron is further equipped with targeting protein, three folate molecules, which lead the DNA nanoparticles to the abundant folate receptors found on some tumors. The result showed that the gene expression targeted by the RNAi, luciferase, dropped by more than half. This study shows promise in using DNA nanotechnology as an effective tool to deliver treatment using the emerging RNA Interference technology.

## Design

DNA nanostructures must be rationally designed so that individual nucleic acid strands will assemble into the desired structures. This process usually begins with specification of a desired target structure or function. Then, the overall secondary structure of the target complex is determined, specifying the arrangement of nucleic acid strands within the structure, and which portions of those strands should be bound to each other. The last step is the primary structure design, which is the specification of the actual base sequences of each nucleic acid strand.

## Structural Design

The first step in designing a nucleic acid nanostructure is to decide how a given structure should be represented by a specific arrangement of nucleic acid strands. This design step determines the secondary structure, or the positions of the base pairs that hold the individual strands together in the desired shape. Several approaches have been demonstrated:

- Tile-based structures. This approach breaks the target structure into smaller units with strong binding between the strands contained in each unit, and weaker interactions between the units. It is often used to make periodic lattices, but can also be used to implement algorithmic self-assembly, making them a platform for DNA computing. This was the dominant design strategy used from the mid-1990s until the mid-2000s, when the DNA origami methodology was developed.

- Folding structures. An alternative to the tile-based approach, folding approaches make the nanostructure from one long strand, which can either have a designed sequence that folds due to its interactions with itself, or it can be folded into the desired shape by using shorter, "staple" strands. This latter method is called DNA origami, which allows forming nanoscale two- and three-dimensional shapes.

- Dynamic assembly. This approach directly controls the kinetics of DNA self-assembly, specifying all of the intermediate steps in the reaction mechanism in addition to the final product. This is done using starting materials which adopt a hairpin structure; these then assemble into the final conformation in a cascade reaction, in a specific order. This approach has the advantage of proceeding isothermally, at a constant temperature. This is in contrast to the thermodynamic approaches, which require a thermal annealing step where a temperature change is required to trigger the assembly and favor proper formation of the desired structure.

## Sequence Design

After any of the above approaches are used to design the secondary structure of a target complex, an actual sequence of nucleotides that will form into the desired structure must be devised. Nucleic acid design is the process of assigning a specific nucleic acid base sequence to each of a structure's constituent strands so that they will associate into a desired conformation. Most methods have the goal of designing sequences so that the target structure has the lowest energy, and is thus the most thermodynamically favorable, while incorrectly assembled structures have higher energies and are thus disfavored. This is done either through simple, faster heuristic methods such as sequence symmetry minimization, or by using a full nearest-neighbor thermodynamic model, which is more accurate but slower and more computationally intensive. Geometric models are used to examine tertiary structure of the nanostructures and to ensure that the complexes are not overly strained.

Nucleic acid design has similar goals to protein design. In both, the sequence of monomers is designed to favor the desired target structure and to disfavor other structures. Nucleic acid design has the advantage of being much computationally easier than protein design, because the simple base pairing rules are sufficient to predict a structure's energetic favorability, and detailed information about the overall three-dimensional folding of the structure is not required. This allows the use of simple heuristic methods that yield experimentally robust designs. However, nucleic acid structures are less versatile than proteins in their function because of proteins' increased ability to fold into complex structures, and the limited chemical diversity of the four nucleotides as compared to the twenty proteinogenic amino acids.

## Materials and Methods

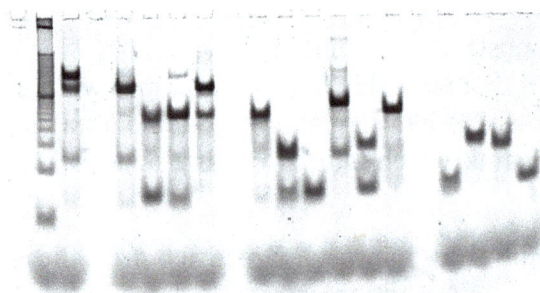

Gel electrophoresis methods, such as this formation assay on a DX complex, are used to ascertain whether the desired structures are forming properly. Each vertical lane contains a series of bands, where each band is characteristic of a particular reaction intermediate.

The sequences of the DNA strands making up a target structure are designed computationally, using molecular modeling and thermodynamic modeling software. The nucleic acids themselves are then synthesized using standard oligonucleotide synthesis methods, usually automated in an oligonucleotide synthesizer, and strands of custom sequences are commercially available. Strands can be purified by denaturing gel electrophoresis if needed, and precise concentrations determined via any of several nucleic acid quantitation methods using ultraviolet absorbance spectroscopy.

The fully formed target structures can be verified using native gel electrophoresis, which gives size and shape information for the nucleic acid complexes. An electrophoretic mobility shift assay can assess whether a structure incorporates all desired strands. Fluorescent labeling and Förster resonance energy transfer (FRET) are sometimes used to characterize the structure of the complexes.

Nucleic acid structures can be directly imaged by atomic force microscopy, which is well suited to extended two-dimensional structures, but less useful for discrete three-dimensional structures because of the microscope tip's interaction with the fragile nucleic acid structure; transmission electron microscopy and cryo-electron microscopy are often used in this case. Extended three-dimensional lattices are analyzed by X-ray crystallography.

## History

The woodcut *Depth* (pictured) by M. C. Escher reportedly inspired Nadrian Seeman to consider using three-dimensional lattices of DNA to orient hard-to-crystallize molecules. This led to the beginning of the field of DNA nanotechnology.

The conceptual foundation for DNA nanotechnology was first laid out by Nadrian Seeman in the early 1980s. Seeman's original motivation was to create a three-dimensional DNA lattice for orienting other large molecules, which would simplify their crystallographic study by eliminating the difficult process of obtaining pure crystals. This idea had reportedly come to him in late 1980, after realizing the similarity between the woodcut *Depth* by M. C. Escher and an array of DNA six-arm junctions. Several natural branched DNA structures were known at the time, including the DNA replication fork and the mobile Holliday junction, but Seeman's insight was that immobile nucleic acid junctions could be created by properly designing the strand sequences to remove symmetry in the assembled molecule, and that these immobile junctions could in principle be combined into rigid crystalline lattices. The first theoretical paper proposing this scheme was published in 1982, and the first experimental demonstration of an immobile DNA junction was published the following year.

In 1991, Seeman's laboratory published a report on the synthesis of a cube made of DNA, the first synthetic three-dimensional nucleic acid nanostructure, for which he received the 1995 Feynman Prize in Nanotechnology. This was followed by a DNA truncated octahedron. However, it soon became clear that these structures, polygonal shapes with flexible junctions as their vertices, were not rigid enough to form extended three-dimensional lattices. Seeman developed the more rigid double-crossover (DX) structural motif, and in 1998, in collaboration with Erik Winfree, published the creation of two-dimensional lattices of DX tiles. These tile-based structures had the advantage that they provided the ability to implement DNA computing, which was demonstrated by Winfree and Paul Rothemund in their 2004 paper on the algorithmic self-assembly of a Sierpinski gasket structure, and for which they shared the 2006 Feynman Prize in Nanotechnology. Winfree's key insight was that the DX tiles could be used as Wang tiles, meaning that their assembly could perform computation. The synthesis of a three-dimensional lattice was finally published by Seeman in 2009, nearly thirty years after he had set out to achieve it.

New abilities continued to be discovered for designed DNA structures throughout the 2000s. The first DNA nanomachine—a motif that changes its structure in response to an input—was demonstrated in 1999 by Seeman. An improved system, which was the first nucleic acid device to make use of toehold-mediated strand displacement, was demonstrated by Bernard Yurke the following year. The next advance was to translate this into mechanical motion, and in 2004 and 2005, several DNA walker systems were demonstrated by the groups of Seeman, Niles Pierce, Andrew Turberfield, and Chengde Mao. The idea of using DNA arrays to template the assembly of other molecules such as nanoparticles and proteins, first suggested by Bruche Robinson and Seeman in 1987, was demonstrated in 2002 by Seeman, Kiehl et al. and subsequently by many other groups.

In 2006, Rothemund first demonstrated the DNA origami method for easily and robustly forming folded DNA structures of arbitrary shape. Rothemund had conceived of this method as being conceptually intermediate between Seeman's DX lattices, which used many short strands, and William Shih's DNA octahedron, which consisted mostly of one very long strand. Rothemund's DNA origami contains a long strand which folding is assisted by several short strands. This method allowed forming much larger structures than formerly possible, and which are less technically demanding to design and synthesize. DNA origami was the cover story of *Nature* on March 15, 2006. Rothemund's research demonstrating two-dimensional DNA origami structures was followed by the demonstration of solid three-dimensional DNA origami by Douglas *et al.* in 2009, while the labs of Jørgen Kjems and Yan demonstrated hollow three-dimensional structures made out of two-dimensional faces.

DNA nanotechnology was initially met with some skepticism due to the unusual non-biological use of nucleic acids as materials for building structures and doing computation, and the preponderance of proof of principle experiments that extended the abilities of the field but were far from actual applications. Seeman's 1991 paper on the synthesis of the

DNA cube was rejected by the journal *Science* after one reviewer praised its originality while another criticized it for its lack of biological relevance. By the early 2010s, however, the field was considered to have increased its abilities to the point that applications for basic science research were beginning to be realized, and practical applications in medicine and other fields were beginning to be considered feasible. The field had grown from very few active laboratories in 2001 to at least 60 in 2010, which increased the talent pool and thus the number of scientific advances in the field during that decade.

# Energy Applications of Nanotechnology

Over the past few decades, the fields of science and engineering have been seeking to develop new and improved types of energy technologies that have the capability of improving life all over the world. In order to make the next leap forward from the current generation of technology, scientists and engineers have been developing energy applications of nanotechnology. Nanotechnology, a new field in science, is any technology that contains components smaller than 100 nanometers. For scale, a single virus particle is about 100 nanometers in width.

An important subfield of nanotechnology related to energy is nanofabrication. Nanofabrication is the process of designing and creating devices on the nanoscale. Creating devices smaller than 100 nanometers opens many doors for the development of new ways to capture, store, and transfer energy. The inherent level of control that nanofabrication could give scientists and engineers would be critical in providing the capability of solving many of the problems that the world is facing today related to the current generation of energy technologies.

People in the fields of science and engineering have already begun developing ways of utilizing nanotechnology for the development of consumer products. Benefits already observed from the design of these products are an increased efficiency of lighting and heating, increased electrical storage capacity, and a decrease in the amount of pollution from the use of energy. Benefits such as these make the investment of capital in the research and development of nanotechnology a top priority.

## Consumer Products

Recently, previously established and entirely new companies such as BetaBatt, Inc. and Oxane Materials are focusing on nanomaterials as a way to develop and improve upon older methods for the capture, transfer, and storage of energy for the development of consumer products.

ConsERV, a product developed by the Dais Analytic Corporation, uses nanoscale polymer membranes to increase the efficiency of heating and cooling systems and has already proven to be a lucrative design. The polymer membrane was specifically configured for this application by selectively engineering the size of the pores in the membrane to prevent air from passing, while allowing moisture to pass through the membrane. ConsERV's value is demonstrated in the form of an energy recovery a device which pretreats the incoming fresh air to a building using the energy found in the exhaust air steam using no moving parts to lower the energy and carbon footprint of existing forms of heating and cooling equipment Polymer membranes can be designed to selectively allow particles of one size and shape to pass through while preventing other. This makes for a powerful tool that can be used in all markets - consumer, commercial, industrial, and government products from biological weapons protection to industrial chemical separations. Dais's near term uses of this 'family' of selectively engineered nanotechnology materials, aside from ConsERV, include (a.) a completely new cooling cycle capable of replacing the refrigerant based cooling cycle the world has known for the past 100 plus years. This product, under development, is named NanoAir. NanoAir uses only water and this selectively engineered membrane material to cool (or heat) and dehumidify (or humidify) air. There are no fluorocarbon producing gasses used, and the energy required to cool a space drops as thermodynamics does the actual cooling. The company was awarded an Advanced Research Program Administration - Energy award in 2010, and a United States Department of Defense (DoD) grant in 2011 both designed to accelerate this newer, energy efficient technology closer to commercialization, and (b.) a novel way to clean most all contaminated forms of water called NanoClear. By using the selectivity of this hermetic, engineered composite material it can transfer only a water molecule from one face of the membrane to the other leaving behind the contaminants. It should also be noted Dais received a US Patent (Patent Number 7,990,679) in October 2011 titled "Nanoparticle Ultracapacitor". This patented item again uses the selectively engineered material to create an energy storage mechanism projected to have performance and cost advantages over existing storage technologies. The company has used this patent's concepts to create a functional energy storage prototype device named NanoCap. NanoCap is a form of ultra-capacitor potentially useful to power a broad range of applications including most forms of transportation, energy storage (especially useful as a storage media for renewable energy technologies), telecommunication infrastructure, transistor gate dielectrics, and consumer battery applications (cell phones, computers, etc.).

A New York-based company called Applied NanoWorks, Inc. has been developing a consumer product that utilizes LED technology to generate light. Light-emitting diodes or LEDs, use only about 10% of the energy that a typical incandescent or fluorescent light bulb uses and typically last much longer, which makes them a viable alternative to traditional light bulbs. While LEDs have been around for decades, this company and others like it have been developing a special variant of LED called the white LED. White LEDs consist of semi-conducting organic layers that are only about 100 nanometers in distance from each other and are placed between two electrodes, which create an anode, and a cathode. When voltage is applied to the system, light is generated when electricity passes through the two organic layers. This is called electroluminescence. The semiconductor properties of the organic layers are what allow for the minimal amount of energy necessary to generate light. In traditional light bulbs, a metal filament is used to generate light when electricity is run through the filament. Using metal generates a great deal of heat and therefore lowers efficiency.

Research for longer lasting batteries has been an ongoing process for years. Researchers have now begun to utilize nanotechnology for battery technology. mPhase Technologies in conglomeration with Rutgers University and Bell Laboratories have utilized nanomaterials to alter the wetting behavior of the surface where the liquid in the battery lies to spread the liquid droplets over a greater area on the surface and therefore have greater control over the movement of the droplets. This gives more control to the designer of the battery. This control prevents reactions in the battery by separating the electrolytic liquid from the anode and the cathode when the battery is not in use and joining them when the battery is in need of use.

Thermal applications also are a future applications of nanothechonlogy creating low cost system of heating, ventilation, and air conditioning, changing molecular structure for better management of temperature

## Reduction of Energy Consumption

A reduction of energy consumption can be reached by better insulation systems, by the use of more efficient lighting or combustion systems, and by use of lighter and stronger materials in the transportation sector. Currently used light bulbs only convert approximately 5% of the electrical energy into light. Nanotechnological approaches like or quantum caged atoms (QCAs) could lead to a strong reduction of energy consumption for illumination.

## Increasing the Efficiency of Energy Production

Today's best solar cells have layers of several different semiconductors stacked together to absorb light at different energies but they still only manage to use 40 percent of the Sun's energy. Commercially available solar cells have much lower efficiencies (15-20%). Nanotechnology could help increase the efficiency of light conversion by using nanostructures with a continuum of bandgaps.

The degree of efficiency of the internal combustion engine is about 30-40% at present. Nanotechnology could improve combustion by designing specific catalysts with maximized surface area. In 2005, scientists at the University of Toronto developed a spray-on nanoparticle substance that, when applied to a surface, instantly transforms it into a solar collector.

## Nuclear Accident Cleanup and Waste Storage

Nanomaterials deployed by swarm robotics may be helpful for decontaminating a site of a nuclear accident which poses hazards to humans because of high levels of radiation and radioactive particles. Hot nuclear compounds such as corium or melting fuel rods may be contained in "bubbles" made from nanomaterials that are designed to isolate the harmful effects of nuclear activity occurring inside of them from the outside environment that organisms inhabit.

## Economic Benefits

The relatively recent shift toward using nanotechnology with respect to the capture, transfer, and storage of energy has and will continue to have many positive economic impacts on society. The control of materials that nanotechnology offers to scientists and engineers of consumer products is one of the most important aspects of nanotechnology. This allows for an improved efficiency of products across the board.

A major issue with current energy generation is the loss of efficiency from the generation of heat as a by-product of the process. A common example of this is the heat generated by the internal combustion engine. The internal combustion engine loses about 64% of the energy from gasoline as heat and an improvement of this alone could have a significant economic impact. However, improving the internal combustion engine in this respect has proven to be extremely difficult without sacrificing performance. Improving the efficiency of fuel cells through the use of nanotechnology appears to be more plausible by using molecularly tailored catalysts, polymer membranes, and improved fuel storage.

In order for a fuel cell to operate, particularly of the hydrogen variant, a noble-metal catalyst (usually platinum, which is very expensive) is needed to separate the electrons from the protons of the hydrogen atoms. However, catalysts of this type are extremely sensitive to carbon monoxide reactions. In order to combat this, alcohols or hydrocarbons compounds are used to lower the carbon monoxide concentration in the system. This adds an additional cost to the device. Using nanotechnology, catalysts can be designed through nanofabrication that are much more resistant to carbon monoxide reactions, which improves the efficiency of the process and may be designed with cheaper materials to additionally lower costs.

Fuel cells that are currently designed for transportation need rapid start-up periods for the practicality of consumer use. This process puts a lot of strain on the traditional polymer electrolyte membranes, which decreases the life of the membrane requiring frequent replacement. Using nanotechnology, engineers have the ability to create a much more durable polymer membrane, which addresses this problem. Nanoscale polymer membranes are also much more efficient in ionic conductivity. This improves the efficiency of the system and decreases the time between replacements, which lowers costs.

Another problem with contemporary fuel cells is the storage of the fuel. In the case of hydrogen fuel cells, storing the hydrogen in gaseous rather than liquid form improves the efficiency by 5%. However, the materials that we currently have available to us significantly limit fuel storage due to low stress tolerance and costs. Scientists have come up with an answer to this by using a nanoporous styrene material (which is a relatively inexpensive material) that when super-cooled to around -196°C, naturally holds on to hydrogen atoms and when heated again releases the hydrogen for use.

## Capacitors: Then and Now

For decades, scientists and engineers have been attempting to make computers smaller and more efficient. A crucial component of computers are capacitors. A capacitor is a device that is made of a pair of electrodes separated by an insulator that each stores an opposite charge. A capacitor stores a charge when it is removed from the circuit that it is connected to; the charge is released when it is replaced back into the circuit. Capacitors have an advantage over batteries in that they release their charge much more quickly than a battery.

Traditional or foil capacitors are composed of thin metal conducting plates separated by an electrical insulator, which are then stacked or rolled and placed in a casing. The problem with a traditional capacitor such as this is that they limit how small an engineer can design a computer. Scientists and engineers have since turned to nanotechnology for a solution to the problem.

Using nanotechnology, researchers developed what they call "ultracapacitors." An ultracapacitor is a general term that describes a capacitor that contains nanocomponents. Ultracapacitors are being researched heavily because of their high density interior, compact size, reliability, and high capacitance. This decrease in size makes it increasingly possible to develop much smaller circuits and computers. Ultracapacitors also have the capability to supplement batteries in hybrid vehicles by providing a large amount of energy during peak acceleration and allowing the battery to supply energy over longer periods of time, such as during a constant driving speed. This could decrease the size and weight of the large batteries needed in hybrid vehicles as well as take additional stress off the battery. However, the combination of ultracapacitors and a battery is not cost effective due to the need of additional DC/DC electronics to coordinate the two.

Nanoporous carbon aerogel is one type of material that is being utilized for the design of ultracapacitors. These aerogels have a very large interior surface area and can have its properties altered by changing the pore diameter and distribution along with adding nanosized alkali metals to alter its conductivity.

Carbon nanotubes are another possible material for use in an ultracapacitor. Carbon nanotubes are created by vaporizing carbon and allowing it to condense on a surface. When the carbon condenses, it forms a nanosized tube composed of carbon atoms. This tube has a high surface area, which increases the amount of charge that can be stored. The low reliability and high cost of using carbon nanotubes for ultracapacitors is currently an issue of research.

In a study concerning ultracapacitors or supercapacitors, researchers at the Sungkyunk-wan University in the Republic of Korea explored the possibility of increasing the capacitance of electrodes through the addition of fluorine atoms to the walls of carbon nanotubes. As briefly mentioned before, carbon nanotubes are an increasing form of capacitors due to their superb chemical stability, high conductivity, light mass, and their large surface area. These researchers fluorinated single-walled carbon nanotubes (SWCNTs) at high temperatures to bind fluorine atoms to the walls. The attached fluorine atoms changed the non-polar nanotubes to become polar molecules. This can be attributed to the charge transfer from the fluorine. This created dipole-dipole layers along the carbon nanotube walls. Testing of these fluorinated SWCNTs against normal state SWCNTs showed a difference in capacitance. It was determined that the fluorinated SWCNTs are advantageous in fabricating electrodes for capacitors and improve the wettability with aqueous electrolytes, which promotes the overall performance of supercapacitors. While this study brought to knowledge a more efficient example of capacitors, little is known about this new supercapacitor, large scale synthesis is lacking and is necessary for any massive production, and preparation conditions are quite tedious in achieving the final product.

## Theory of Capacitance

Understanding the concept of capacitance can be helpful in understanding why nanotechnology is such a powerful tool for the design of higher energy storing capacitors. A capacitor's capacitance (C) or amount of energy stored is equal to the amount of charge (Q) stored on each plate divided by the voltage (V) between the plates. Another representation of capacitance is that capacitance (C) is approximately equal to the permittivity ($\varepsilon$) of the dielectric times the area (A) of the plates divided by the distance (d) between them. Therefore, capacitance is proportional to the surface area of the conducting plate and inversely proportional to the distance between the plates.

Using carbon nanotubes as an example, a property of carbon nanotubes is that they have a very high surface area to store a charge. Using the above proportionality that capacitance (C) is proportional to the surface area (A) of the conducting plate; it becomes obvious that using nanoscaled materials with high surface area would be great for increasing capacitance. The other proportionality described above is that capacitance (C) is inversely proportional to the distance (d) between the plates. Using nanoscaled plates such as carbon nanotubes with nanofabrication techniques, gives the capability of decreasing the space between plates which again increases capacitance.

## Industrial Applications of Nanotechnology

Nanotechnology is impacting the field of consumer goods, several products that incorporate nanomaterials are already in a variety of items; many of which people do not even realize contain nanoparticles, products with novel functions ranging from easy-to-clean to scratch-resistant. Examples of that car bumpers are made lighter, clothing is more stain repellant, sunscreen is more radiation resistant, synthetic bones are stronger, cell phone screens are lighter weight, glass packaging for drinks leads to a longer shelf-life, and balls for various sports are made more durable. Using nanotech, in the mid-term modern textiles will become "smart", through embedded "wearable electronics", such novel products have also a promising potential especially in the field of cosmetics, and has numerous potential applications in heavy industry. Nanotechnology is predicted to be a main driver of technology and business in this century and holds the promise of higher performance materials, intelligent systems and new production methods with significant impact for all aspects of society.

## Foods

A complex set of engineering and scientific challenges in the food and bioprocessing industry for manufacturing high quality and safe food through efficient and sustainable means can be solved through nanotechnology. Bacteria identification and food quality monitoring using biosensors; intelligent, active, and smart food packaging systems; nanoencapsulation of bioactive food compounds are few examples of emerging applications of nanotechnology for the food industry. Nanotechnology can be applied in the production, processing, safety and packaging of food. A nanocomposite coating process could improve food packaging by placing anti-microbial agents directly on the surface of the coated film. Nanocomposites could increase or decrease gas permeability of different fillers as is needed for different products. They can also improve the mechanical and heat-resistance properties and lower the oxygen transmission rate. Research is being performed to apply nanotechnology to the detection of chemical and biological substances for sensanges in foods.

In general, food substances are not allowed to be adulterated, according to the Food, Drug and Cosmetic Act (section 402). Additives to food must conform to all regulations in the food additives amendment of 1958 as well as the FDA Modernization Act of 1997. In addition, color additives are obliged to comply with all regulations stipulated by the Color Additive Amendments of 1960. A safety assessment must be performed on all food substances for submission and approval by the US FDA. The mandatory information in this assessment includes the identity, technical effects, self-limiting levels of use, dietary exposure and safety studies for the manufacturing processes used, including the use of nanotechnology. Food manufacturers are obliged to assess whether the identity, safety or regulatory status of a food substance is affected by significant changes in manufacturing processes, such as the use of nanotechnology. In their guidance document published in April 2012, the US FDA discusses what considerations and recommendations may apply to such an assessment.

## Nano-foods

New foods are among the nanotechnology-created consumer products coming onto the market at the rate of 3 to 4 per week, according to the Project on Emerging Nanotechnologies (PEN), based on an inventory it has drawn up of 609 known or claimed nano-products. On PEN's list are three foods—a brand of canola cooking oil called Canola Active Oil, a tea called Nanotea and a chocolate diet shake called Nanoceuticals Slim Shake Chocolate. According to company information posted on PEN's Web site, the canola oil, by Shemen Industries of Israel, contains an additive called "nanodrops" designed to carry vitamins, minerals and phytochemicals through the digestive system and urea. The shake, according to U.S. manufacturer RBC Life Sciences Inc., uses cocoa infused "NanoClusters" to enhance the taste and health benefits of cocoa without the need for extra sugar.

## Consumer Goods

### Surfaces and Coatings

The most prominent application of nanotechnology in the household is self-cleaning or "easy-to-clean" surfaces on ceramics or glasses. Nano ceramic particles have improved the smoothness and heat resistance of common household equipment such as the flat iron.

The first sunglasses using protective and anti-reflective ultrathin polymer coatings are on the market. For optics, nanotechnology also offers scratch resistant surface coatings based on nanocomposites. Nano-optics could allow for an increase in precision of pupil repair and other types of laser eye surgery.

### Textiles

The use of engineered nanofibers already makes clothes water- and stain-repellent or wrinkle-free. Textiles with a nanotechnological finish can be washed less frequently and at lower temperatures. Nanotechnology has been used to integrate tiny carbon particles membrane and guarantee full-surface protection from electrostatic charges for the wearer. Many other applications have been developed by research institutions such as the Textiles Nanotechnology Laboratory at Cornell University, and the UK's Dstl and its spin out company P2i.

### Cosmetics

One field of application is in sunscreens. The traditional chemical UV protection approach suffers from its poor long-term stability. A sunscreen based on mineral nanoparticles such as titanium oxide offer several advantages. Titanium oxide nanoparticles have a comparable UV protection property as the bulk material, but lose the cosmetically undesirable whitening as the particle size is decreased.

### Sports

Nanotechnology may also play a role in sports such as soccer, football, and baseball.

Materials for new athletic shoes may be made in order to make the shoe lighter (and the athlete faster). Baseball bats already on the market are made with carbon nanotubes that reinforce the resin, which is said to improve its performance by making it lighter. Other items such as sport towels, yoga mats, exercise mats are on the market and used by players in the National Football League, which use antimicrobial nanotechnology to prevent parasuram from illnesses caused by bacteria such as Methicillin-resistant Staphylococcus aureus (commonly known as MRSA).

## Aerospace and Vehicle Manufacturers

Lighter and stronger materials will be of immense use to aircraft manufacturers, leading to increased performance. Spacecraft will also benefit, where weight is a major factor. Nanotechnology might thus help to reduce the size of equipment and thereby decrease fuel-consumption required to get it airborne. Hang gliders may be able to halve their weight while increasing their strength and toughness through the use of nanotech materials. Nanotech is lowering the mass of supercapacitors that will increasingly be used to give power to assistive electrical motors for launching hang gliders off flatland to thermal-chasing altitudes.

Much like aerospace, lighter and stronger materials would be useful for creating vehicles that are both faster and safer. Combustion engines might also benefit from parts that are more hard-wearing and more heat-resistant.

## Military

## Biological Sensors

Nanotechnology can improve the military's ability to detect biological agents. By using nanotechnology, the military would be able to create sensor systems that could detect biological agents. The sensor systems are already well developed and will be one of the first forms of nanotechnology that the military will start to use.

## Uniform Material

Nanoparticles can be injected into the material on soldiers' uniforms to not only make the material more durable, but also to protect soldiers from many different dangers such as high temperatures, impacts and chemicals. The nanoparticles in the material protect soldiers from these dangers by grouping together when something strikes the armor and stiffening the area of impact. This stiffness helps lessen the impact of whatever hit the armor, whether it was extreme heat or a blunt force. By reducing the force of the impact, the nanoparticles protect the soldier wearing the uniform from any injury the impact could have caused.

Another way nanotechnology can improve soldiers' uniforms is by creating a better form of camouflage. Mobile pigment nanoparticles injected into the material can produce a better form of camouflage. These mobile pigment particles would be able to change the color of the uniforms depending upon the area that the soldiers are in. There is still much research being done on this self-changing camouflage.

Nanotechnology can improve thermal camouflage. Thermal camouflage helps protect soldiers from people who are using night vision technology. Surfaces of many different military items can be designed in a way that electromagnetic radiation can help lower the infrared signatures of the object that the surface is on. Surfaces of soldiers' uniforms and surfaces of military vehicle are a few surfaces that can be designed in this way. By lowering the infrared signature of both the soldiers and the military vehicles the soldiers are using, it will provide better protection from infrared guided weapons or infrared surveillance sensors.

## Communication Method

There is a way to use nanoparticles to create coated polymer threads that can be woven into soldiers' uniforms. These polymer threads could be used as a form of communication between the soldiers. The system of threads in the uniforms could be set to different light wavelengths, eliminating the ability for anyone else to listen in. This would lower the risk of having anything intercepted by unwanted listeners.

## Medical System

A medical surveillance system for soldiers to wear can be made using nanotechnology. This system would be able to watch over their health and stress levels. The systems would be able to react to medical situations by releasing drugs or compressing wounds as necessary. This means that if the system detected an injury that was bleeding, it would be able to compress around the wound until further medical treatment could be received. The system would also be able to release drugs into the soldier's body for health reasons, such as pain killers for an injury. The system would be able to inform the medics at base of the soldier's health status at all times that the soldier is wearing the system. The energy needed to communicate this information back to base would be produced through the soldier's body movements.

## Weapons

Nanoweapon is the name given to military technology currently under development which seeks to exploit the power of nanotechnology in the modern battlefield.

## Risks in Military

- People such as state agencies, criminals and enterprises could use nano-robots to eavesdrop on conversations held in private.

- Grey goo: an uncontrollable, self-replicating nano-machine or robot.

- Nanoparticles used in different military materials could potentially be a hazard to the soldiers that are wearing the material, if the material is allowed to get worn out. As the uniforms wear down it is possible for nanomaterial to break off and enter the soldiers' bodies. Having nanoparticles entering the soldiers' bodies would be very unhealthy and could seriously harm them. There is not a lot of information on what the actual damage to the soldiers would be, but there have been studies on the effect of nanoparticles entering a fish through its skin. The studies showed that the different fish in the study suffered from varying degrees of brain damage. Although brain damage would be a serious negative effect, the studies also say that the results cannot be taken as an accurate example of what would happen to soldiers if nanoparticles entered their bodies. There are very strict regulations on the scientists that manufacture products with nanoparticles. With these strict regulations, they are able to largely decrease the danger of nanoparticles wearing off of materials and entering the soldiers' systems.

## Catalysis

Chemical catalysis benefits especially from nanoparticles, due to the extremely large surface to volume ratio. The application potential of nanoparticles in catalysis ranges from fuel cell to catalytic converters and photocatalytic devices. Catalysis is also important for the production of chemicals. For example, nanoparticles with a distinct chemical surrounding (ligands), or specific optical properties.

Platinum nanoparticle are now being considered in the next generation of automotive catalytic converters because the very high surface area of nanoparticles could reduce the amount of platinum required. However, some concerns have been raised due to experiments demonstrating that they will spontaneously combust if methane is mixed with the ambient air. Ongoing research at the Centre National de la Recherche Scientifique (CNRS) in France may resolve their true usefulness for catalytic applications. Nanofiltration may come to be an important application, although future research must be careful to investigate possible toxicity.

## Construction

Nanotechnology has the potential to make construction faster, cheaper, safer, and more varied. Automation of nanotechnology construction can allow for the creation of structures from advanced homes to massive skyscrapers much more quickly and at much lower cost. In the near future, Nanotechnology can be used to sense cracks in foundations of architecture and can send nanobots to repair them.

Nanotechnology is an active research area that encompasses a number of disciplines such as electronics, bio-mechanics and coatings. These disciplines assist in the areas of civil engineering and construction materials. If nanotechnology is implemented in the construction of homes and infrastructure, such structures will be stronger. If buildings are stronger, then fewer of them will require reconstruction and less waste will be produced.

Nanotechnology in construction involves using nanoparticles such as alumina and silica. Manufacturers are also investigating the methods of producing nano-cement. If cement with nano-size particles can be manufactured and processed, it will open up a large number of opportunities in the fields of ceramics, high strength composites and electronic applications.

Nanomaterials still have a high cost relative to conventional materials, meaning that they are not likely to feature in high-volume building materials. The day when this technology slashes the consumption of structural steel has not yet been contemplated.

## Cement

Much analysis of concrete is being done at the nano-level in order to understand its structure. Such analysis uses various techniques developed for study at that scale such as Atomic Force Microscopy (AFM), Scanning Electron Microscopy (SEM) and Focused Ion Beam (FIB). This has come about as a side benefit of the development of these instruments to study the nanoscale in general, but the understanding of the structure and behavior of concrete at the fundamental level is an important and very appropriate use of nanotechnology. One of the fundamental aspects of nanotechnology is its interdisciplinary nature and there has already been cross over research between the mechanical modeling of bones for medical engineering to that of concrete which has enabled the study of chloride diffusion in concrete (which causes corrosion of reinforcement). Concrete is, after all, a macro-material strongly influenced by its nano-properties and understanding it at this new level is yielding new avenues for improvement of strength, durability and monitoring as outlined in the following paragraphs.

Silica ($SiO_2$) is present in conventional concrete as part of the normal mix. However, one of the advancements made by the study of concrete at the nanoscale is that particle packing in concrete can be improved by using nano-silica which leads to a densifying of the micro and nanostructure resulting in improved mechanical properties. Nano-silica addition to cement based materials can also control the degradation of the fundamental C-S-H (calcium-silicatehydrate) reaction of concrete caused by calcium leaching in water as well as block water penetration and therefore lead to improvements in durability. Related to improved particle packing, high energy milling of ordinary Portland cement (OPC) clinker and standard sand, produces a greater particle size diminution with respect to conventional OPC and, as a result, the compressive strength of the refined material is also 3 to 6 times higher (at different ages).

## Steel

Steel is a widely available material that has a major role in the construction industry. The use of nanotechnology in steel helps to improve the physical properties of steel. Fatigue, or the structural failure of steel, is due to cyclic loading. Current steel designs are based on the reduction in the allowable stress, service life or regular inspection regime. This has a significant impact on the life-cycle costs of structures and limits the effective use of resources. Stress risers are responsible for initiating cracks from which fatigue failure results. The addition of copper nanoparticles reduces the surface un-evenness of steel, which then limits the number of stress risers and hence fatigue cracking. Advancements in this technology through the use of nanoparticles would lead to increased safety, less need for regular inspection, and more efficient materials free from fatigue issues for construction.

Steel cables can be strengthened using carbon nanotubes. Stronger cables reduce the costs and period of construction, especially in suspension bridges, as the cables are run from end to end of the span.

The use of vanadium and molybdenum nanoparticles improves the delayed fracture problems associated with high strength bolts. This reduces the effects of hydrogen embrittlement and improves steel micro-structure by reducing the effects of the inter-granular cementite phase.

Welds and the Heat Affected Zone (HAZ) adjacent to welds can be brittle and fail without warning when subjected to sudden dynamic loading. The addition of nanoparticles such as magnesium and calcium makes the HAZ grains finer in plate steel. This nanoparticle addition leads to an increase in weld strength. The increase in strength results in a smaller resource requirement because less material is required in order to keep stresses within allowable limits.

## Wood

Nanotechnology represents a major opportunity for the wood industry to develop new products, substantially reduce processing costs, and open new markets for biobased materials.

Wood is also composed of nanotubes or "nanofibrils"; namely, lignocellulosic (woody tissue) elements which are twice as strong as steel. Harvesting these nanofibrils would lead to a new paradigm in sustainable construction as both the production and use would be part of a renewable cycle. Some developers have speculated that building functionality onto lignocellulosic surfaces at the nanoscale could open new opportunities for such things as self-sterilizing surfaces, internal self-repair, and electronic lignocellulosic devices. These non-obtrusive active or passive nanoscale sensors would provide feedback on product performance and environmental conditions during service by monitoring structural loads, temperatures, moisture content, decay fungi, heat losses or gains, and loss of conditioned air. Currently, however, research in these areas appears limited.

Due to its natural origins, wood is leading the way in cross-disciplinary research and modelling techniques. BASF have developed a highly water repellent coating based on the actions of the lotus leaf as a result of the incorporation of silica and alumina nanoparticles and hydrophobic polymers. Mechanical studies of bones have been adapted to model wood, for instance in the drying process.

## Glass

Research is being carried out on the application of nanotechnology to glass, another important material in construction. Titanium dioxide ($TiO_2$) nanoparticles are used to coat glazing since it has sterilizing and anti-fouling properties. The particles catalyze powerful reactions that break down organic pollutants, volatile organic compounds and bacterial membranes. $TiO_2$ is hydrophilic (attraction to water), which can attract rain drops that then wash off the dirt particles. Thus the introduction of nanotechnology in the Glass industry, incorporates the self-cleaning property of glass.

Fire-protective glass is another application of nanotechnology. This is achieved by using a clear intumescent layer sandwiched between glass panels (an interlayer) formed of silica nanoparticles ($SiO_2$), which turns into a rigid and opaque fire shield when heated. Most of glass in construction is on the exterior surface of buildings. So the light and heat entering the building through glass has to be prevented. The nanotechnology can provide a better solution to block light and heat coming through windows.

## Coatings

Coatings is an important area in construction coatings are extensively use to paint the walls, doors, and windows. Coatings should provide a protective layer bound to the base material to produce a surface of the desired protective or functional properties. The coatings should have self healing capabilities through a process of "self-assembly". Nanotechnology is being applied to paints to obtained the coatings having self healing capabilities and corrosion protection under insulation. Since these coatings are hydrophobic and repels water from the metal pipe and can also protect metal from salt water attack.

Nanoparticle based systems can provide better adhesion and transparency. The $TiO_2$ coating captures and breaks down organic and inorganic air pollutants by a photocatalytic process, which leads to putting roads to good environmental use.

## Fire Protection and Detection

Fire resistance of steel structures is often provided by a coating produced by a spray-on-cementitious process. The nano-cement has the potential to create a new paradigm in this area of application because the resulting material can be used as a tough, durable, high temperature coating. It provides a good method of increasing fire resistance and this is a cheaper option than conventional insulation.

## Risks in Construction

In building construction nanomaterials are widely used from self-cleaning windows to flexible solar panels to wi-fi blocking paint. The self-healing concrete, materials to block ultraviolet and infrared radiation, smog-eating coatings and light-emitting walls and ceilings are the new nanomaterials in construction. Nanotechnology is a promise for making the "smart home" a reality. Nanotech-enabled sensors can monitor temperature, humidity, and airborne toxins, which needs nanotech-based improved batteries. The building components will be intelligent and interactive since the sensor uses wireless components, it can collect the wide range of data.

If nanosensors and nanomaterials become an everyday part of the buildings, as with smart homes, what are the consequences of these materials on human beings?

1.  Effect of nanoparticles on health and environment: Nanoparticles may also enter the body if building water supplies are filtered through commercially available nanofilters. Airborne and waterborne nanoparticles enter from building ventilation and wastewater systems.

2.  Effect of nanoparticles on societal issues: As sensors become commonplace, a loss of privacy and autonomy may result from users interacting with increasingly intelligent building components.

## Nanoremediation

Nanoremediation is the use of nanoparticles for environmental remediation. It is being explored to treat ground water, wastewater, soil, sediment, or other contaminated environmental materials. Nanoremediation is an emerging industry; by 2009, nanoremediation technologies had been documented in at least 44 cleanup sites around the world, predominantly in the United States. In Europe, nanoremediation is being investigated by the EC funded NanoRem Project. A report produced by the NanoRem consortium has identified around 70 nanoremediation projects worldwide at pilot or full scale. During nanoremediation, a nanoparticle agent must be brought into contact with the target contaminant under conditions that allow a detoxifying or immobilizing reaction. This process typically involves a pump-and-treat process or *in situ* application.

Some nanoremediation methods, particularly the use of nano zero-valent iron for groundwater cleanup, have been deployed at full-scale cleanup sites. Other methods remain in research phases.

## Applications

Nanoremediation has been most widely used for groundwater treatment, with additional extensive research in wastewater treatment. Nanoremediation has also been tested for soil and sediment cleanup. Even more preliminary research is exploring the use of nanoparticles to remove toxic materials from gases.

## Groundwater Remediation

Currently, groundwater remediation is the most common commercial application of nanoremediation technologies. Using nanomaterials, especially zero-valent metals (ZVMs), for groundwater remediation is an emerging approach that is promising due to the availability and effectiveness of many nanomaterials for degrading or sequestering contaminants.

Nanotechnology offers the potential to effectively treat contaminants *in situ*, avoiding excavation or the need to pump contaminated water out of the ground. The process begins with nanoparticles being injected into a contaminated aquifer via an injection well. The nanoparticles are then transported by groundwater flow to the source of contamination. Upon contact, nanoparticles can sequester contaminants (via adsorption or complexation), immobilizing them, or they can degrade the contaminants to less harmful compounds. Contaminant transformations are typically redox reactions. When the nanoparticle is the oxidant or reductant, it is considered reactive.

The ability to inject nanoparticles to the subsurface and transport them to the contaminant source is imperative for successful treatment. Reactive nanoparticles can be injected into a well where they will then be transported down gradient to the contaminated area. Drilling and packing a well is quite expensive. Direct push wells cost less than drilled wells and are the most often used delivery tool for remediation with nanoiron. A nanoparticle slurry can be injected along the vertical range of the probe to provide treatment to specific aquifer regions.

## Surface Water Treatment

The use of various nanomaterials, including carbon nanotubes and $TiO_2$, shows promise for treatment of surface water, including for purification, disinfection, and desalination. Target contaminants in surface waters include heavy metals, organic contaminants, and pathogens. In this context, nanoparticles may be used as sorbents, as reactive agents (photocatalysts or redox agents), or in membranes used for nanofiltration.

## Trace Contaminant Detection

Nanoparticles may assist in detecting trace levels of contaminants in field settings, contributing to effective remediation. Instruments that can operate outside of a laboratory often are not sensitive enough to detect trace contaminants. Rapid, portable, and cost-effective measurement systems for trace contaminants in groundwater and other environmental media would thus enhance contaminant detection and cleanup. One potential method is to separate the analyte from the sample and concentrate them to a smaller volume, easing detection and measurement. When small quantities of solid sorbents are used to absorb the target for concentration, this method is referred to as solid-phase microextraction.

With their high reactivity and large surface area, nanoparticles may be effective sorbents to help concentrate target contaminants for solid-phase microextraction, particularly in the form of self-assembled monolayers on mesoporous supports. The mesoporous silica structure, made through a surfactant templated sol-gel process, gives these self-assembled monolayers high surface area and a rigid open pore structure. This material may be an effective sorbent for many targets, including heavy metals such as mercury, lead, and cadmium, chromate and arsenate, and radionuclides such as $^{99}$Tc, $^{137}$CS, uranium, and the actinides.

## Mechanism

The small size of nanoparticles leads to several characteristics that may enhance remediation. Nanomaterials are highly reactive because of their high surface area per unit mass. Their small particle size also allows nanoparticles to enter small pores in soil or sediment that larger particles might not penetrate, granting them access to contaminants sorbed to soil and increasing the likelihood of contact with the target contaminant.

Because nanomaterials are so tiny, their movement is largely governed by Brownian motion as compared to gravity. Thus, the flow of groundwater can be sufficient to transport the particles. Nanoparticles then can remain suspended in solution longer to establish an *in situ* treatment zone.

Once a nanoparticle contacts the contaminant, it may degrade the contaminant, typically through a redox reaction, or adsorb to the contaminant to immobilize it. In some cases, such as with magnetic nano-iron, adsorbed complexes may be separated from the treated substrate, removing the contaminant. Target contaminants include organic molecules such as pesticides or organic solvents and metals such as arsenic or lead. Some research is also exploring the use of nanoparticles to remove excessive nutrients such as nitrogen and phosphorus.

## Materials

A variety of compounds, including some that are used as macro-sized particles for remediation, are being studied for use in nanoremediation. These materials include zero-valent metals like zero-valent iron, calcium carbonate, carbon-based compounds such as graphene or carbon nanotubes, and metal oxides such as titanium dioxide and iron oxide.

## Nano Zero-valent Iron

As of 2012, nano zero-valent iron (nZVI) was the nanoscale material most commonly used in bench and field remediation tests. nZVI may be mixed or coated with another metal, such as palladium, silver, or copper, that acts as a catalyst in what is called a bimetallic nanoparticle. nZVI may also be emulsified with a surfactant and an oil, creating a membrane that enhances the nanoparticle's ability to interact with hydrophobic liquids and protects it against reactions with materials dissolved in water. Commercial nZVI particle sizes may sometimes exceed true "nano" dimensions (100 nm or less in diameter).

nZVI appears to be useful for degrading organic contaminants, including chlorinated organic compounds such as polychlorinated biphenyls (PCBs) and trichloroethene (TCE), as well as immobilizing or removing metals. nZVI and other nanoparticles that do not require light can be injected belowground into the contaminated zone for *in situ* groundwater remediation and, potentially, soil remediation.

nZVI nanoparticles can be prepared by using sodium borohydride as the key reductant. $NaBH_4$ (0.2 M) is added into $FeCl_3 \cdot 6H_2$ (0.05 M) solution (~1:1 volume ratio). Ferric iron is reduced via the following reaction:

$$4Fe^{3+} + 3BH_4^- + 9H_2O \rightarrow 4Fe^0 + 3H_2BO_3^- + 12H^+ + 6H_2$$

Palladized Fe particles are prepared by soaking the nanoscale iron particles with an ethanol solution of 1wt% of palladium acetate ($[Pd(C_2H_3O_2)2]_3$). This causes the reduction and deposition of Pd on the Fe surface:

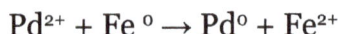

$$Pd^{2+} + Fe^0 \rightarrow Pd^0 + Fe^{2+}$$

Similar methods may be used to prepared Fe/Pt, Fe/Ag, Fe/Ni, Fe/Co, and Fe/Cu bimetallic particles. With the above methods, nanoparticles of diameter 50-70 nm may be produced. The average specific surface area of Pd/Fe particles is about 35 $m^2$/g. Ferrous iron salt has also been successfully used as the precursor.

## Titanium Dioxide

Titanium dioxide ($TiO_2$) is also a leading candidate for nanoremediation and wastewater treatment, although as of 2010 it is reported to have not yet been expanded to full-scale commercialization. When exposed to ultraviolet light, such as in sunlight, titanium dioxide produces hydroxyl radicals, which are highly reactive and can oxidize contaminants. Hydroxyl radicals are used for water treatment in methods generally termed advanced oxidation processes. Because light is required for this reaction, $TiO_2$ is not appropriate for underground *in situ* remediation, but it may be used for wastewater treatment or pump-and-treat groundwater remediation.

$TiO_2$ is inexpensive, chemically stable, and insoluble in water. $TiO_2$ has a wide band gap energy (3.2 eV) that requires the use of UV light, as opposed to visible light only, for photocatalytic activation. To enhance the efficiency of its photocatalysis, research has investigated modifications to $TiO_2$ or alternative photocatalysts that might use a greater portion of photons in the visible light spectrum. Potential modifications include doping $TiO_2$ with metals, nitrogen, or carbon.

## Challenges

When using *in situ* remediation the reactive products must be considered for two reasons. One reason is that a reactive product might be more harmful or mobile than the parent compound. Another reason is that the products can affect the effectiveness and/ or cost of remediation. TCE (trichloroethylene), under reducing conditions by nanoiron, may sequentially dechlorinate to DCE (dichloroethene) and VC (vinyl chloride). VC is known to be more harmful than TCE, meaning this process would be undesirable.

Nanoparticles also react with non-target compounds. Bare nanoparticles tend to clump together and also react rapidly with soil, sediment, or other material in ground water. For *in situ* remediation, this action inhibits the particles from dispersing in the contaminated area, reducing their effectiveness for remediation. Coatings or other treatment may allow nanoparticles to disperse farther and potentially reach a greater portion of the contaminated zone. Coatings for nZVI include surfactants, polyelectrolyte coatings, emulsification layers, and protective shells made from silica or carbon.

Such designs may also affect the nanoparticles' ability to react with contaminants, their uptake by organisms, and their toxicity. A continuing area of research involves the potential for nanoparticles used for remediation to disperse widely and harm wildlife, plants, or people.

In some cases, bioremediation may be used deliberately at the same site or with the same material as nanoremediation. Ongoing research is investigating how nanoparticles may interact with simultaneous biological remediation.

## Nanofiltration

Water desalination

Methods

- Distillation

    o   Multi-stage flash distillation (MSF)

    o   Multiple-effect distillation (MED|ME)

    o   Vapor-compression (VC)

- Ion exchange

- Membrane processes

    o   Electrodialysis reversal (EDR)

    o   Reverse osmosis (RO)

    o   Nanofiltration (NF)

    o   Membrane distillation (MD)

    o   Forward osmosis (FO)

- Freezing desalination

- Geothermal desalination

- Solar desalination

    o   Solar humidification–dehumidification (HDH)

    o   Multiple-effect humidification (MEH)

- Methane hydrate crystallization

- High grade water recycling

- Seawater greenhouse

---

Nanofiltration is a relatively recent membrane filtration process used most often with low total dissolved solids water such as surface water and fresh groundwater, with the purpose of softening (polyvalent cation removal) and removal of disinfection by-product precursors such as natural organic matter and synthetic organic matter.

Nanofiltration is also becoming more widely used in food processing applications such as dairy, for simultaneous concentration and partial (monovalent ion) demineralisation.

## General

Nanofiltration is a membrane filtration-based method that uses nanometer sized cylindrical through-pores that pass through the membrane at 90°. Nanofiltration membranes have pore sizes from 1-10 nanometers, smaller than that used in microfiltration and ultrafiltration, but just larger than that in reverse osmosis. Membranes used are predominantly created from polymer thin films. Materials that are commonly use include polyethylene terephthalate or metals such as aluminum. Pore dimensions are controlled by pH, temperature and time during development with pore densities ranging from 1 to 106 pores per cm². Membranes made from polyethylene terephthalate and other similar materials, are referred to as "track-etch" membranes, named after the way the pores on the membranes are made. "Tracking" involves bombarding the polymer thin film with high energy particles. This results in making tracks that are chemically developed into the membrane, or "etched" into the membrane, which are the pores. Membranes created from metal such as alumina membranes, are made by electrochemically growing a thin layer of aluminum oxide from aluminum metal in an acidic medium.

## Range of Applications

Historically, nanofiltration and other membrane technology used for molecular separation was applied entirely on aqueous systems. The original uses for nanofiltration were water treatment and in particular water softening. Nanofilters can "soften" water by retaining scale-forming, hydrated divalent ions (e.g. $Ca^{2+}$, $Mg^{2+}$) while passing smaller hydrated monovalent ions .

In recent years, the use of nanofiltration has been extended into other industries such as milk and juice production. Research and development in solvent-stable membranes has allowed the application for nanofiltration membranes to extend into new areas such as pharmaceuticals, fine chemicals, and flavour and fragrance industries. Development in organic solvent nanofiltration technology and commercialization of membranes used has extended possibilities for applications in a variety of organic solvents ranging from non-polar through polar to polar aprotic.

| Industry | Uses |
| --- | --- |
| Fine chemistry and Pharmaceuticals | Non-thermal solvent recovery and management<br>Room temperature solvent exchange |
| Oil and Petroleum chemistry | Removal of tar components in feed<br>Purification of gas condensates |
| Bulk Chemistry | Product Polishing<br>Continuous recovery of homogeneous catalysts |
| Natural Essential Oils and similar products | Fractionation of crude extracts<br>Enrichment of natural compounds Gentle Separations |
| Medicine | Able to extract amino acids and lipids from blood and other cell culture. |

## Advantages and Disadvantages

One of the main advantages of nanofiltration as a method of softening water is that during the process of retaining calcium and magnesium ions while passing smaller hydrated monovalent ions, filtration is performed without adding extra sodium ions, as used in ion exchangers. Many separation processes do not operate at room temperature (e.g. distillation), which greatly increases the cost of the process when continuous heating or cooling is applied. Performing gentle molecular separation is linked with nanofiltration that is often not included with other forms of separation processes (centrifugation). These are two of the main benefits that are associated with nanofiltration. Nanofiltration has a very favorable benefit of being able to process large volumes and continuously produce streams of have a Nanofiltration is the least used method of membrane filtration in industry as the membrane pores sizes are limited to only nanometers. Anything smaller, reverse osmosis is used and anything larger is used for ultrafiltration. Ultrafiltration can also be used in cases where nanofiltration can be used, due to it being more conventional. A main disadvantage associated with nanotechnology, as with all membrane filter technology, is the cost and maintenance of the membranes used. Nanofiltration membranes are an expensive part of the process. Repairs and replacement of membranes is dependent on total dissolved solids, flow rate and components of the feed. With nanofiltration being used across various industries, only an estimation of replacement frequency can be used. This causes nanofilters to be replaced a short time before or after their prime usage is complete.

## Design and Operation

Industrial applications of membranes require hundreds to thousands of square meters of membranes and therefore an efficient way to reduce the footprint by packing them is required. Membranes first became commercially viable when low cost methods of housing in 'modules' were achieved. Membranes are not self-supporting. They need to be stayed by a porous support that can withstand the pressures required to operate the NF membrane without hindering the performance of the membrane. To do this effectively, the module needs to provide a channel to remove the membrane permeation and provide appropriate flow condition that reduces the phenomena of concentration polarisation. A good design minimises pressure losses on both the feed side and permeate side and thus energy requirements. Leakage of the feed into the permeate stream must also be prevented. This can be done through either the use of permanent seals such as glue or replaceable seals such as O-rings.

## Concentration Polarisation

Concentration polarisation describes the accumulation of the species being retained close to the surface of the membrane which reduces separation capabilities. It occurs because the particles are convected towards the membrane with the solvent and its magnitude is the balance between this convection caused by solvent flux and the particle transport away from the membrane due to the concentration gradient (predominantly caused by diffusion.) Although concentration polarisation is easily reversible, it can lead to fouling of the membrane.

## Spiral Wound Module

Spiral wound modules are the most commonly used style of module and are 'standardized' design, available in a range of standard diameters (2.5", 4" and 8") to fit standard pressure vessel that can hold several modules in series connected by O-rings. The module uses flat sheets wrapped around a central tube. The membranes are glued along three edges over a permeate spacer to form 'leaves'. The permeate spacer supports the membrane and conducts the permeate to the central permeate tube. Between each leaf, a mesh like feed spacer is inserted. The reason for the mesh like dimension of the spacer is to provide a hydrodynamic environment near the surface of the membrane that discourages concentration polarisation. Once the leaves have been wound around the central tube, the module is wrapped in a casing layer and caps placed on the end of the cylinder to prevent 'telescoping' that can occur in high flow rate and pressure conditions.

## Tubular Module

Tubular modules look similar to shell and tube heat exchangers with bundles of tubes with the active surface of the membrane on the inside. Flow through the tubes is normally turbulent, ensuring low concentration polarisation but also increasing energy costs. The tubes can either be self-supporting or supported by insertion into perforated metal tubes. This module design is limited for nanofiltration by the pressure they can withstand before bursting, limiting the maximum flux possible. Due to both the high energy operating costs of turbulent flow and the limiting burst pressure, tubular modules are more suited to 'dirty' applications where feeds have particulates such as filtering raw water to gain potable water in the Fyne process. The membranes can be easily cleaned through a 'pigging' technique with foam balls are squeezed through the tubes, scouring the caked deposits.

## Flux Enhancing Strategies

These strategies work to reduce the magnitude of concentration polarisation and fouling. There is a range of techniques available however the most common is feed channel spacers as described in spiral wound modules. All of the strategies work by increasing eddies and generating a high shear in the flow near the membrane surface. Some of these strategies include vibrating the membrane, rotating the membrane, having a rotor disk above the membrane, pulsing the feed flow rate and introducing gas bubbling close to the surface of the membrane.

## Characterisation

Many different factors must be taken into account in the design of NF membranes, since they vary so much in material, separation mechanisms, morphology and thus application. Two important parameters should be investigated during preliminary calculations, performance and morphology parameters.

## Performance Parameters

Retention of both charged and uncharged solutes and permeation measurements can be categorised into performance parameters since the performance under natural conditions of a membrane is based on the ratio of solute retained/ permeated through the membrane.

For charged solutes, the ionic distribution of salts near the membrane-solution interface plays an important role in determining the retention characteristic of a membrane. If the charge of the membrane and the composition and concentration of the solution to be filtered is known, the distribution of various salts can be found. This in turn can be combined with the known charge of the membrane and the Gibbs–Donnan effect to predict the retention characteristics for that membrane.

Uncharged solutes cannot be characterised simply by Molecular Weight Cut Off (MWCO,) although in general an increase in molecular weight or solute size leads to an increase in retention. The chemical structure, functional end-groups as well as pH of the solute, all play an important role in determining the retention characteristics and as such detailed information about the solute molecule characteristics must be known before implementing a NF design.

## Morphology Parameters

The morphology of a membrane must also be known in order to implement a successful design of a NF system, and this is usually done by microscopy. Atomic force microscopy (AFM) is one method used to characterise the surface roughness of a membrane by passing a small sharp tip (<100 Å) across the surface of a membrane and measuring the resulting Van der Waals force between the atoms in the end of the tip and the surface. This is useful as a direct correlation between surface roughness and colloidal fouling has been developed. Correlations also exist between fouling and other morphology parameters, such as hydrophobe, showing that the more hydrophobic a membrane is, the less prone to fouling it is.

Methods to determine the porosity of porous membranes have also been found via permporometry, making use of differing vapour pressures to characterise the pore size and pore size distribution within the membrane. Initially all pores in the membrane are completely filled with a liquid and as such no permeation of a gas occurs, but after reducing the relative vapour pressure some gaps will start to form within the pores as dictated by the Kelvin equation. Polymeric (non-porous) membranes cannot be subjected to this methodology as the condensable vapour should have a negligible interaction within the membrane.

## Typical Figures for Industrial Applications

Keeping in mind that NF is usually part of a composite system for purification, a single unit is chosen based off the design specifications for the NF unit. For drinking water purification many commercial membranes exist, coming from different chemical families, having different structures, chemical tolerances and salt rejections and so the characterisation must be chosen based on the chemical composition and concentration of the feed stream.

NF units in drinking water purification range from extremely low salt rejection (<5% in 1001A membranes) to almost complete rejection (99% in 8040-TS80-TSA membranes.) Flow rates range from 25–60 m³/day for each unit, so commercial filtration requires multiple NF units in parallel to process large quantities of feed water. The pressures required in these units are generally between 4.5-7.5 bar.

For seawater desalination using a NF-RO system a typical process is shown below.

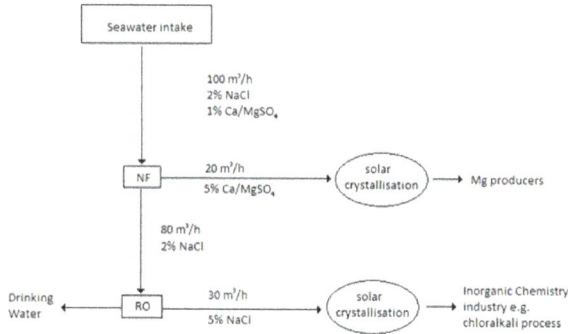

Because of the fact that NF permeate is rarely clean enough to be used as the final product for drinking water and other water purification, is it commonly used as a pre treatment step for reverse osmosis (RO) as is shown above.

## Post Treatment

As with other membrane based separations such as ultrafiltration, microfiltration and reverse osmosis, post-treatment of eitherpermeate or retentate flow streams (depending on the application) – is a necessary stage in industrial NF separation prior to commercial distribution of the product. The choice and order of unit operations employed in post-treatment is dependent on water quality regulations and the design of the NF system. Typical NF water purification post-treatment stages include aeration and disinfection & stabilisation.

## Aeration

A Polyvinyl chloride (PVC) or fibre-reinforced plastic (FRP) degasifier is used to remove dissolved gases such as carbon dioxide and hydrogen sulfide from the permeate stream. This is achieved by blowing air in a countercurrent direction to the water falling through packing material in the degasifier. The air effectively strips the unwanted gases from the water.

## Disinfection & Stabilisation

The permeate water from a NF separation is demineralised and may be disposed to large changes in pH, thus providing a substantial risk of corrosion in piping and other equipment components. To increase the stability of the water, chemical addition of alkaline solutions such as lime and caustic soda is employed. Furthermore, disinfectants such as chlorine or chloroamine are added to the permeate, as well as phosphate or fluoride corrosion inhibitors in some cases.

## New Developments

Contemporary research in the area of Nanofiltration (NF) technology is primarily concerned with improving the performance of NF membranes, minimising membrane fouling and reducing energy requirements of already existing processes. One way in which researchers are attempting to improve NF performance – more specifically increase permeate flux and lower membrane resistance – is through experimentation with different membrane materials and configurations. thin film composite membranes (TFC), which consist of a number of extremely thin selective layers interfacially polymerized over a microporous substrate, have had the most commercial success in industrial membrane applications due to the capability of optimizing the selectivity and permeability of each individual layer. Recent research has shown that the addition of nanotechnology materials such as electrospunnanofibrous membrane layers (ENMs) to conventional TFC membranes results in an enhanced permeate flux. This has been attributed to inherent properties of ENMs that favour flux, namely their interconnected pore structure, high porosity and low transmembrane pressure. A recently developed membrane configuration which offers a more energy efficient alternative to the commonly used spiral wound arrangement is the hollow fibre membrane. This format has the advantage of requiring significantly less pre-treatment than spiral wound membranes, as solids introduced in the feed are displaced effectively during backwash or flushing. As a result, membrane fouling and pre-treatment energy costs are reduced. Extensive research has also been conducted on the potential use of Titanium Dioxide ($TiO_2$, titania) nanoparticles for membrane fouling reduction. This method involves applying a nonporous coating of titania onto the membrane surface. Internal fouling/pore blockage of the membrane is resisted due to the nonporosity of the coating, whilst the superhydrophilic nature of titania provides resistance to surface fouling by reducing adhesion of emulsified oil on the membrane surface.

## References

- Hall, J. Storrs (2005). Nanofuture : what's next for nanotechnology. Amherst, NY: Prometheus Books. ISBN 978-1591022879.

- Chan WCW (2007). "Toxicity Studies of Fullerenes and Derivatives". Bio-applications of nanoparticles. New York, NY: Springer Science + Business Media. ISBN 0-387-76712-6.

- Petty, M.C.; Bryce, M.R.; Bloor, D. (1995). An Introduction to Molecular Electronics. London: Edward Arnold. ISBN 0-19-521156-1.

- Waldner, Jean-Baptiste (2007). Nanocomputers and Swarm Intelligence. London: ISTE. p. 26. ISBN 1-84704-002-0.

- Hanft, Susan (2011). Market Research Report Nanotechnology in water treatment. Wellesley, MA USA: BCC Research. p. 16. ISBN 1596237090.

- Raymond D. Letterman (ed.)(1999). "Water Quality and Treatment." 5th Ed. (New York: American Water Works Association and McGraw-Hill.) ISBN 0-07-001659-3.

- Background: Pelesko, John A. (2007). Self-assembly: the science of things that put themselves together. New York: Chapman & Hall/CRC. pp. 5, 7. ISBN 978-1-58488-687-7.

- Applications: Rietman, Edward A. (2001). Molecular engineering of nanosystems. Springer. pp. 209–212. ISBN 978-0-387-98988-4. Retrieved 17 April 2011.

- History: Pelesko, John A. (2007). Self-assembly: the science of things that put themselves together. New York: Chapman & Hall/CRC. pp. 201, 242, 259. ISBN 978-1-58488-687-7.

- Raymond D. Letterman (ed.)(1999). "Water Quality and Treatment." 5th Ed. (New York: American Water Works Association and McGraw-Hill.) ISBN 0-07-001659-3.

- American Water Works Association (2007). Manual of Water Supply Practices in Reverse Osmosis and Nanofiltration. Denver: American Water Works Association. pp. 101–102. ISBN 1583214917.

- Cloete, TE et al (editor) (2010). Nanotechnology in Water Treatment Applications. Caister Academic Press. ISBN 978-1-904455-66-0.

- "Market report on emerging nanotechnology now available". Market Report. US National Science Foundation. 25 February 2014. Retrieved 7 June 2016.

- "Recent progress on DNA based walkers". Current Opinion in Biotechnology. 34: 56–64. doi:10.1016/j.copbio.2014.11.017. Retrieved 2015-09-28.

- Theron, J.; J. A. Walker; T. E. Cloete (2008-01-01). "Nanotechnology and Water Treatment: Applications and Emerging Opportunities". Critical Reviews in Microbiology. 34 (1): 43–69. doi:10.1080/10408410701710442. ISSN 1040-841X. Retrieved 2014-07-29.

- Chong, Meng Nan; Bo Jin; Christopher W. K. Chow; Chris Saint (May 2010). "Recent developments in photocatalytic water treatment technology: A review". Water Research. 44 (10): 2997–3027. doi:10.1016/j.watres.2010.02.039. ISSN 0043-1354. Retrieved 2014-07-29.

# 6

# Allied Fields of Nanoengineering

Nanoengineering is a vast subject that has a number of allied fields that have been thoroughly discussed in this section. Some of the fields explained in the following text are nanophotonics, materials science, molecular engineering, tissue engineering, ceramic engineering and nanometrology. The study of the behavior of light on the scale of nanometer is known as nanohotonics and the discovery and designing of new materials is referred to as materials science and engineering. In order to completely understand nanoengineering, it is necessary to understand the fields allied to it.

## Nanoarchitectonics

Nanoarchitectonics is a scientific jargon term coined at the National Institute for Materials Science for one of its leading units, International Center for Materials Nanoarchitectonics (MANA). It refers to a technology allowing to arrange nanoscale structural units, which are usually a group of atoms or molecules, in an intended configuration.

Nanoarchitectonics is further classified into two topics, "Nano Creation" and "Nano Organization". "Nano Creation" is synthesis of a new material that does not exist in Nature. For example, by peeling atomic sheets off graphite slab, a novel nano-material graphene can be obtained, which has very different properties from graphite.

A typical example of the "Nano Organization" is the development of a nanoelectronics circuit. Challenging electronic devices are produced experimentally, using previously discovered materials, such as carbon nanotubes, fullerenes, graphene, single molecules having functional groups, etc. However, their practical use is impossible without a technology ("Nano Organization") to integrate and link these devices into a system.

Nanoarchitectonics is not limited to "Nano Creation" and "Nano Organization", but rather employs those techniques to understand and use the ultimate functions of materials. The important technologies to achieve this goal involve manipulation of single atoms and molecules through physical interactions, chemical reactions, applied fields or self-assembly.

Examples of those technologies are the following:

- Physical manipulation of atoms and molecules has already been achieved using, e.g., atomically sharp needles of a scanning tunneling microscope or an atomic force microscope.

- Chemical manipulation can be realized through, e.g., electrochemical reactions induced by localized electric field in a nanoelectronic device or through local polymerization.

- Field-induced manipulation is a widely explored direction where control over atoms or molecules is achieved using various combinations of electric, magnetic, elastic and other fields. A well-known example is manipulation of individual atoms by laser beams ("optical tweezers").

- Self-assembly usually involves weaker interactions, such as van der Waals forces. Common self-assembly examples are growth of a molecular crystal, e.g., snow.

# Nanophotonics

Nanophotonics or Nano-optics is the study of the behavior of light on the nanometer scale, and of the interaction of nanometer-scale objects with light. It is a branch of optics, optical engineering, electrical engineering, and nanotechnology. It often (but not exclusively) involves metallic components, which can transport and focus light via surface plasmon polaritons.

The term "nano-optics", just like the term "optics", usually concerns ultraviolet, visible, and near-infrared light (free-space wavelength around 300-1200 nanometers).

## Background

Normal optical components, like lenses and microscopes, generally cannot normally focus light to nanometer (deep subwavelength) scales, because of the diffraction limit (Rayleigh criterion). Nevertheless, it is possible to squeeze light into a nanometer scale using other techniques like, for example, surface plasmons, localized surface plasmons around nanoscale metal objects, and the nanoscale apertures and nanoscale sharp tips used in near-field scanning optical microscopy (NSOM) and photoassisted scanning tunnelling microscopy.

## Motivations

Nanophotonics researchers pursue a very wide variety of goals, in fields ranging from biochemistry to electrical engineering. A few of these goals are summarized below.

### Optoelectronics and Microelectronics

If light can be squeezed into a small volume, it can be absorbed and detected by a small detector. Small photodetectors tend to have a variety of desirable properties including low noise, high speed, and low voltage and power.

Small lasers have various desirable properties for optical communication including low threshold current (which helps power efficiency) and fast modulation (which means more data transmission). Very small lasers require subwavelength optical cavities. An example is spasers, the surface plasmon version of lasers.

Integrated circuits are made using photolithography, i.e. exposure to light. In order to make very small transistors, the light needs to be focused into extremely sharp images. Using various techniques such as immersion lithography and phase-shifting photomasks, it has indeed been possible to make images much finer than the wavelength—for example, drawing 30 nm lines using 193 nm light. Plasmonic techniques have also been proposed for this application.

Heat-assisted magnetic recording is a nanophotonic approach to increasing the amount of data that a magnetic disk drive can store. It requires a laser to heat a tiny, subwavelength area of the magnetic material before writing data. The magnetic write-head would have metal optical components to concentrate light at the right location.

Miniaturization in optoelectronics, for example the miniaturization of transistors in integrated circuits, has improved their speed and cost. However, optoelectronic circuits can only be miniaturized if the optical components are shrunk along with the electronic components. This is relevant for on-chip optical communication (i.e. passing information from one part of a microchip to another by sending light through optical waveguides, instead of changing the voltage on a wire).

## Solar Cells

Solar cells often work best when the light is absorbed very close to the surface, both because electrons near the surface have a better chance of being collected, and because the device can be made thinner, which reduces cost. Researchers have investigated a variety of nanophotonic techniques to intensify light in the optimal locations within a solar cell.

## Spectroscopy

*Using nanophotonics to create high peak intensities*: If a given amount of light energy is squeezed into a smaller and smaller volume ("hot-spot"), the intensity in the hot-spot gets larger and larger. This is especially helpful in nonlinear optics; an example is surface enhanced Raman scattering. It also allows sensitive spectroscopy measurements of even single molecules located in the hot-spot, unlike traditional spectroscopy methods which take an average over millions or billions of molecules.

## Microscopy

One goal of nanophotonics is to construct a so-called "superlens", which would use metamaterials or other techniques to create images that are more accurate than the diffraction limit (deep subwavelength).

Near-field scanning optical microscope (NSOM or SNOM) is a quite different nanophotonic technique that accomplishes the same goal of taking images with resolution far smaller than the wavelength. It involves raster-scanning a very sharp tip or very small aperture over the surface to be imaged.

Near-field microscopy refers more generally to any technique using the near-field to achieve nanoscale, subwavelength resolution. For example, dual polarization interferometry has picometer resolution in the vertical plane above the wave-guide surface.

## Principles

### Plasmons and Metal Optics

Metals are an effective way to confine light to far below the wavelength. This was originally used in radio and microwave engineering, where metal antennas and waveguides may be hundreds of times smaller than the free-space wavelength. For a similar reason, visible light can be confined to the nano-scale via nano-sized metal structures, such as nano-sized structures, tips, gaps, etc. This effect is somewhat analogous to a lightning rod, where the field concentrates at the tip.

This effect is fundamentally based on the fact that the permittivity of the metal is very large and negative. At very high frequencies (near and above the plasma frequency, usually ultraviolet), the permittivity of a metal is not so large, and the metal stops being useful for concentrating fields.

Many nano-optics designs look like common microwave or radiowave circuits, but shrunk down by a factor of 100,000 or more. After all, radiowaves, microwaves, and visible light are all electromagnetic radiation; they differ only in frequency. So other things equal, a microwave circuit shrunk down by a factor of 100,000 will behave the same way but at 100,000 times higher frequency. For example, researchers have made nano-optical Yagi-Uda antennas following essentially the same design as used for radio Yagi-Uda antennas. Metallic parallel-plate waveguides (striplines), lumped-constant circuit elements such as inductance and capacitance (at visible light frequencies, the values of the latter being of the order of femtohenries and attofarads, respectively), and impedance-matching of dipole antennas to transmission lines, all familiar techniques at microwave frequencies, are some current areas of nanophotonics development. That said, there are a number of very important differences between nano-optics and scaled-down microwave circuits. For example, at optical frequency, metals behave much less like ideal conductors, and also exhibit interesting plasmon-related effects like kinetic inductance and surface plasmon resonance. Likewise, optical fields interact with semi-conductors in a fundamentally different way than microwaves do.

## Near-field Optics

If you take the Fourier transform of an object, it consists of different spatial frequencies. The higher frequencies correspond to the very fine features and sharp edges.

When light is emitted by such an object, the light with very high spatial frequency forms an evanescent wave, which only exists in the near field (very close to the object, within a wavelength or two) and disappears in the far field. This is the origin of the diffraction limit, which says that when a lens images an object, the subwavelength information is blurred out.

Nano-photonics is primarily concerned with the near-field evanescent waves. For example, a superlens (mentioned above) would prevent the decay of the evanescent wave, allowing higher-resolution imaging.

## Metamaterials

Metamaterials are artificial materials engineered to have properties that may not be found in nature. They are created by fabricating an array of structures much smaller than a wavelength. The small (nano) size of the structures is important: That way, light interacts with them as if they made up a uniform, continuous medium, rather than scattering off the individual structures.

## Materials Science

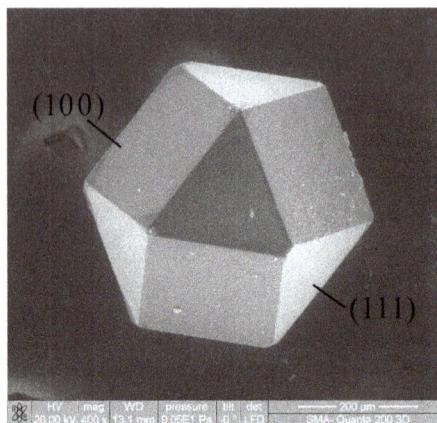

A diamond cuboctahedron showing seven crystallographic planes, imaged with scanning electron microscopy.

The interdisciplinarity field of materials science, also commonly termed materials science and engineering, involves the discovery and design of new materials, with an emphasis on solids. The intellectual origins of materials science stem from the Enlightenment, when researchers began to use analytical thinking from chemistry, physics, and engineering to understand ancient, phenomenological observations in metallurgy and mineralogy. Materials science still incorporates elements of physics, chemistry, and engineering. As such, the field was long considered by academic institutions as a sub-field of these related fields. Beginning in the 1940s, materials science began to be more widely recognized as a specific and distinct field of science and engineering, and major technical universities around the world created dedicated schools of the study.

> Materials science is a syncretic discipline hybridizing metallurgy, ceramics, solid-state physics, and chemistry. It is the first example of a new academic discipline emerging by fusion rather than fission.

Many of the most pressing scientific problems humans currently face are due to the limits of the materials that are available. Thus, breakthroughs in materials science are likely to affect the future of technology significantly.

Materials scientists emphasize understanding how the history of a material (its *processing*) influences its structure, and thus the material's properties and performance. The understanding of processing-structure-properties relationships is called the § materials paradigm. This paradigm is used to advance understanding in a variety of research areas, including nanotechnology, biomaterials, and metallurgy. Materials science is also an important part of forensic engineering and failure analysis - investigating materials, products, structures or components which fail or which do not operate or function as intended, causing personal injury or damage to property. Such investigations are key to understanding, for example, the causes of various aviation accidents and incidents.

## History

A late Bronze Age sword or dagger blade.

The material of choice of a given era is often a defining point. Phrases such as Stone Age, Bronze Age, Iron Age, and Steel Age are great examples. Originally deriving from the manufacture of ceramics and its putative derivative metallurgy, materials science is one of the oldest forms of engineering and applied science. Modern materials science evolved directly from metallurgy, which itself evolved from mining and (likely) ceramics and the use of fire. A major breakthrough in the understanding of materials occurred in the late 19th century, when the American scientist Josiah Willard Gibbs demonstrated that the thermodynamic properties related to atomic structure in various phases are related to the physical properties of a material. Important elements of modern materials science are a product of the space race: the understanding and engineering of the metallic alloys, and silica and carbon materials, used in building space vehicles enabling the exploration of space. Materials science has driven, and been driven by, the development of revolutionary technologies such as rubbers, plastics, semiconductors, and biomaterials.

Before the 1960s (and in some cases decades after), many *materials science* departments were named *metallurgy* departments, reflecting the 19th and early 20th century emphasis on metals. The growth of materials science in the United States was catalyzed in part by the Advanced Research Projects Agency, which funded a series of university-hosted laboratories in the early 1960s "to expand the national program of basic research and training in the materials sciences." The field has since broadened to include every class of materials, including ceramics, polymers, semiconductors, magnetic materials, medical implant materials, biological materials, and nanomaterials.

## Fundamentals

The materials paradigm represented in the form of a tetrahedron.

A material is defined as a substance (most often a solid, but other condensed phases can be included) that is intended to be used for certain applications. There are a myriad of materials around us—they can be found in anything from buildings to spacecraft. Materials can generally be divided into two classes: crystalline and non-crystalline. The traditional examples of materials are metals, semiconductors, ceramics and polymers. New and advanced materials that are being developed include nanomaterials and biomaterials, etc.

The basis of materials science involves studying the structure of materials, and relating them to their properties. Once a materials scientist knows about this structure-property correlation, they can then go on to study the relative performance of a material in a given application. The major determinants of the structure of a material and thus of its properties are its constituent chemical elements and the way in which it has been processed into its final form. These characteristics, taken together and related through the laws of thermodynamics and kinetics, govern a material's microstructure, and thus its properties.

## Structure

As mentioned above, structure is one of the most important components of the field of materials science. Materials science examines the structure of materials from the atomic scale, all the way up to the macro scale. Characterization is the way materials scientists examine the structure of a material. This involves methods such as diffraction with X-rays, electrons, or neutrons, and various forms of spectroscopy and chemical analysis such as Raman spectroscopy, energy-dispersive spectroscopy (EDS), chromatography, thermal analysis, electron microscope analysis, etc. Structure is studied at various levels, as detailed below.

## Atomic Structure

This deals with the atoms of the materials, and how they are arranged to give molecules, crystals, etc. Much of the electrical, magnetic and chemical properties of materials arise from this level of structure. The length scales involved are in angstroms. The way in which the atoms and molecules are bonded and arranged is fundamental to studying the properties and behavior of any material.

## Nanostructure

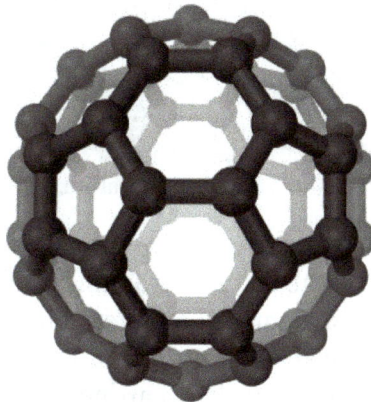

Buckminsterfullerene nanostructure.

Nanostructure deals with objects and structures that are in the 1—100 nm range. In many materials, atoms or molecules agglomerate together to form objects at the nanoscale. This causes many interesting electrical, magnetic, optical, and mechanical properties.

In describing nanostructures it is necessary to differentiate between the number of dimensions on the nanoscale. Nanotextured surfaces have *one dimension* on the nanoscale, i.e., only the thickness of the surface of an object is between 0.1 and 100 nm. Nanotubes have *two dimensions* on the nanoscale, i.e., the diameter of the tube is between 0.1 and 100 nm; its length could be much greater. Finally, spherical nanoparticles have *three dimensions* on the nanoscale, i.e., the particle is between 0.1 and 100 nm in each spatial dimension. The terms nanoparticles and ultrafine particles (UFP) often are used synonymously although UFP can reach into the micrometre range. The term 'nanostructure' is often used when referring to magnetic technology. Nanoscale structure in biology is often called ultrastructure.

Materials which atoms and molecules form constituents in the nanoscale (i.e., they form nanostructure) are called nanomaterials. Nanomaterials are subject of intense research in the materials science community due to the unique properties that they exhibit.

## Microstructure

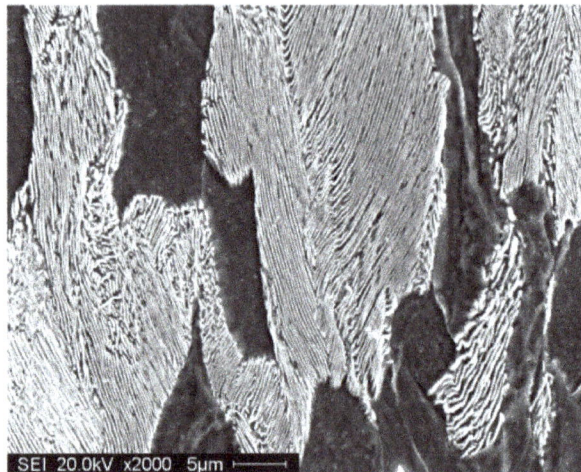

Microstructure of pearlite.

Microstructure is defined as the structure of a prepared surface or thin foil of material as revealed by a microscope above 25× magnification. It deals with objects from 100 nm to a few cm. The microstructure of a material (which can be broadly classified into metallic, polymeric, ceramic and composite) can strongly influence physical properties such as strength, toughness, ductility, hardness, corrosion resistance, high/low temperature behavior, wear resistance, and so on. Most of the traditional materials (such as metals and ceramics) are microstructured.

The manufacture of a perfect crystal of a material is physically impossible. For example, a crystalline material will contain defects such as precipitates, grain boundaries (Hall–Petch relationship), interstitial atoms, vacancies or substitutional atoms. The microstructure of materials reveals these defects, so that they can be studied.

## Macro Structure

Macro structure is the appearance of a material in the scale millimeters to meters—it is the structure of the material as seen with the naked eye.

# Crystallography

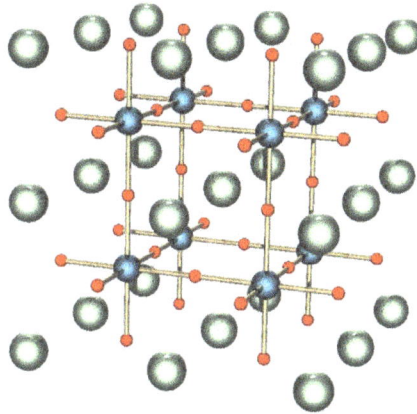

Crystal structure of a perovskite with a chemical formula $ABX_3$.

Crystallography is the science that examines the arrangement of atoms in crystalline solids. Crystallography is a useful tool for materials scientists. In single crystals, the effects of the crystalline arrangement of atoms is often easy to see macroscopically, because the natural shapes of crystals reflect the atomic structure. Further, physical properties are often controlled by crystalline defects. The understanding of crystal structures is an important prerequisite for understanding crystallographic defects. Mostly, materials do not occur as a single crystal, but in polycrystalline form, i.e., as an aggregate of small crystals with different orientations. Because of this, the powder diffraction method, which uses diffraction patterns of polycrystalline samples with a large number of crystals, plays an important role in structural determination. Most materials have a crystalline structure, but some important materials do not exhibit regular crystal structure. Polymers display varying degrees of crystallinity, and many are completely noncrystalline. Glass, some ceramics, and many natural materials are amorphous, not possessing any long-range order in their atomic arrangements. The study of polymers combines elements of chemical and statistical thermodynamics to give thermodynamic and mechanical, descriptions of physical properties.

# Bonding

To obtain a full understanding of the material structure and how it relates to its properties, the materials scientist must study how the different atoms, ions and molecules are arranged and bonded to each other. This involves the study and use of quantum chemistry or quantum physics. Solid-state physics, solid-state chemistry and physical chemistry are also involved in the study of bonding and structure.

# Properties

Materials exhibit myriad properties, including the following.

- Mechanical properties

- Chemical properties

- Electrical properties

- Thermal properties

- Optical properties

- Magnetic properties

The properties of a material determine its usability and hence its engineering application.

## Synthesis and Processing

Synthesis and processing involves the creation of a material with the desired micro-nanostructure. From an engineering standpoint, a material cannot be used in industry if no economical production method for it has been developed. Thus, the processing of materials is vital to the field of materials science.

Different materials require different processing or synthesis methods. For example, the processing of metals has historically been very important and is studied under the branch of materials science named *physical metallurgy*. Also, chemical and physical methods are also used to synthesize other materials such as polymers, ceramics, thin films, etc. As of the early 21st century, new methods are being developed to synthesize nanomaterials such as graphene.

## Thermodynamics

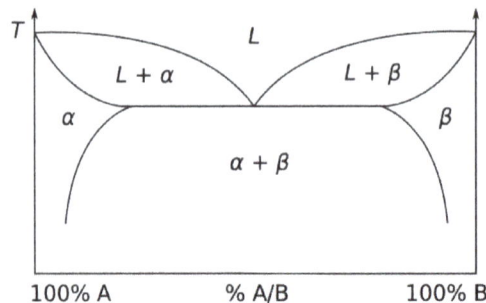

A phase diagram for a binary system displaying a eutectic point.

Thermodynamics is concerned with heat and temperature and their relation to energy and work. It defines macroscopic variables, such as internal energy, entropy, and pressure, that partly describe a body of matter or radiation. It states that the behavior of those variables is subject to general constraints, that are common to all materials, not the peculiar properties of particular materials. These general constraints are expressed in the four laws of thermodynamics. Thermodynamics describes the bulk behavior of the body, not the microscopic behaviors of the very large numbers of its microscopic constituents, such as molecules. The behavior of these microscopic particles is described by, and the laws of thermodynamics are derived from, statistical mechanics.

The study of thermodynamics is fundamental to materials science. It forms the foundation to treat general phenomena in materials science and engineering, including chemical reactions, magnetism, polarizability, and elasticity. It also helps in the understanding of phase diagrams and phase equilibrium.

## Kinetics

Chemical kinetics is the study of the rates at which systems that are out of equilibrium change under the influence of various forces. When applied to materials science, it deals with how a material changes with time (moves from non-equilibrium to equilibrium state) due to application of a certain field. It details the rate of various processes evolving in materials including shape, size, composition and structure. Diffusion is important in the study of kinetics as this is the most common mechanism by which materials undergo change.

Kinetics is essential in processing of materials because, among other things, it details how the microstructure changes with application of heat.

## In Research

Materials science has received much attention from researchers. In most universities, many departments ranging from physics to chemistry to chemical engineering, along with materials science departments, are involved in materials research. Research in materials science is vibrant and consists of many avenues. The following list is in no way exhaustive. It serves only to highlight certain important research areas.

## Nanomaterials

Nanomaterials describe, in principle, materials of which a single unit is sized (in at least one dimension) between 1 and 1000 nanometers ($10^{-9}$ meter) but is usually 1—100 nm.

Nanomaterials research takes a materials science-based approach to nanotechnology, leveraging advances in materials metrology and synthesis which have been developed in support of microfabrication research. Materials with structure at the nanoscale often have unique optical, electronic, or mechanical properties.

The field of nanomaterials is loosely organized, like the traditional field of chemistry, into organic (carbon-based) nanomaterials such as fullerenes, and inorganic nanomaterials based on other elements, such as silicon. Examples of nanomaterials include fullerenes, carbon nanotubes, nanocrystals, etc.

# Biomaterials

The iridescent nacre inside a nautilus shell.

A biomaterial is any matter, surface, or construct that interacts with biological systems. As a science, *bio materials* is about fifty years old. The study of biomaterials is called *bio materials science*. It has experienced steady and strong growth over its history, with many companies investing large amounts of money into developing new products. Biomaterials science encompasses elements of medicine, biology, chemistry, tissue engineering, and materials science.

Biomaterials can be derived either from nature or synthesized in a laboratory using a variety of chemical approaches using metallic components, polymers, ceramics, or composite materials. They are often used and/or adapted for a medical application, and thus comprises whole or part of a living structure or biomedical device which performs, augments, or replaces a natural function. Such functions may be benign, like being used for a heart valve, or may be bioactive with a more interactive functionality such as hydroxylapatite coated hip implants. Biomaterials are also used everyday in dental applications, surgery, and drug delivery. For example, a construct with impregnated pharmaceutical products can be placed into the body, which permits the prolonged release of a drug over an extended period of time. A biomaterial may also be an autograft, allograft or xenograft used as an organ transplant material.

# Electronic, Optical, and Magnetic

Negative index metamaterial.

Semiconductors, metals, and ceramics are used today to form highly complex systems, such as integrated electronic circuits, optoelectronic devices, and magnetic and optical mass storage media. These materials form the basis of our modern computing world, and hence research into these materials is of vital importance.

Semiconductors are a traditional example of these types of materials. They are materials that have properties that are intermediate between conductors and insulators. Their electrical conductivities are very sensitive to impurity concentrations, and this allows for the use of doping to achieve desirable electronic properties. Hence, semiconductors form the basis of the traditional computer.

This field also includes new areas of research such as superconducting materials, spintronics, metamaterials, etc. The study of these materials involves knowledge of materials science and solid-state physics or condensed matter physics.

## Computational Science and Theory

With the increase in computing power, simulating the behavior of materials has become possible. This enables materials scientists to discover properties of materials formerly unknown, as well as to design new materials. Up until now, new materials were found by time consuming trial and error processes. But, now it is hoped that computational methods could drastically reduce that time, and allow tailoring materials properties. This involves simulating materials at all length scales, using methods such as density functional theory, molecular dynamics, etc.

## In Industry

Radical materials advances can drive the creation of new products or even new industries, but stable industries also employ materials scientists to make incremental improvements and troubleshoot issues with currently used materials. Industrial applications of materials science include materials design, cost-benefit tradeoffs in industrial production of materials, processing methods (casting, rolling, welding, ion implantation, crystal growth, thin-film deposition, sintering, glassblowing, etc.), and analytic methods (characterization methods such as electron microscopy, X-ray diffraction, calorimetry, nuclear microscopy (HEFIB), Rutherford backscattering, neutron diffraction, small-angle X-ray scattering (SAXS), etc.).

Besides material characterization, the material scientist or engineer also deals with extracting materials and convering them into useful forms. Thus ingot casting, foundry methods, blast furnace extraction, and electrolytic extraction are all part of the required knowledge of a materials engineer. Often the presence, absence, or variation of minute quantities of secondary elements and compounds in a bulk material will greatly affect the final properties of the materials produced. For example, steels are classified based on 1/10 and 1/100 weight percentages of the carbon and other alloying elements they contain. Thus, the extracting and purifying methods used to extract iron in a blast furnace can affect the quality of steel that is produced.

## Ceramics and Glasses

Si$_3$N$_4$ ceramic bearing parts

Another application of material science is the structures of ceramics and glass, typically associated with the most brittle materials. Bonding in ceramics and glasses uses covalent and ionic-covalent types with SiO$_2$ (silica or sand) as a fundamental building block. Ceramics are as soft as clay or as hard as stone and concrete. Usually, they are crystalline in form. Most glasses contain a metal oxide fused with silica. At high temperatures used to prepare glass, the material is a viscous liquid. The structure of glass forms into an amorphous state upon cooling. Windowpanes and eyeglasses are important examples. Fibers of glass are also available. Scratch resistant Corning Gorilla Glass is a well-known example of the application of materials science to drastically improve the properties of common components. Diamond and carbon in its graphite form are considered to be ceramics.

Engineering ceramics are known for their stiffness and stability under high temperatures, compression and electrical stress. Alumina, silicon carbide, and tungsten carbide are made from a fine powder of their constituents in a process of sintering with a binder. Hot pressing provides higher density material. Chemical vapor deposition can place a film of a ceramic on another material. Cermets are ceramic particles containing some metals. The wear resistance of tools is derived from cemented carbides with the metal phase of cobalt and nickel typically added to modify properties.

## Composites

Filaments are commonly used for reinforcement in composite materials.

Another application of materials science in industry is making composite materials. These are structured materials composed of two or more macroscopic phases. Applications range from structural elements such as steel-reinforced concrete, to the thermal insulating tiles which play a key and integral role in NASA's Space Shuttle thermal protection system which is used to protect the surface of the shuttle from the heat of re-entry into the Earth's atmosphere. One example is reinforced Carbon-Carbon (RCC), the light gray material which withstands re-entry temperatures up to 1,510 °C (2,750 °F) and protects the Space Shuttle's wing leading edges and nose cap. RCC is a laminated composite material made from graphite rayon cloth and impregnated with a phenolic resin. After curing at high temperature in an autoclave, the laminate is pyrolized to convert the resin to carbon, impregnated with furfural alcohol in a vacuum chamber, and cured-pyrolized to convert the furfural alcohol to carbon. To provide oxidation resistance for reuse ability, the outer layers of the RCC are converted to silicon carbide.

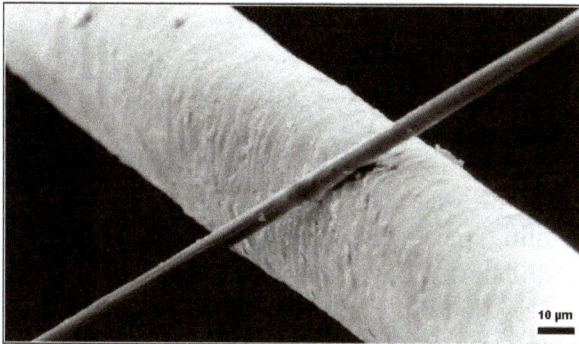

A 6 μm diameter carbon filament (running from bottom left to top right) siting atop the much larger human hair.

Other examples can be seen in the "plastic" casings of television sets, cell-phones and so on. These plastic casings are usually a composite material made up of a thermoplastic matrix such as acrylonitrile butadiene styrene (ABS) in which calcium carbonate chalk, talc, glass fibers or carbon fibers have been added for added strength, bulk, or electrostatic dispersion. These additions may be termed reinforcing fibers, or dispersants, depending on their purpose.

## Polymers

Polymers are chemical compounds made up of a large number of identical components linked together like chains. They are an important part of materials science. Polymers are the raw materials (the resins) used to make what are commonly called plastics and rubber. Plastics and rubber are really the final product, created after one or more polymers or additives have been added to a resin during processing, which is then shaped into a final form. Plastics which have been around, and which are in current widespread use, include polyethylene, polypropylene, polyvinyl chloride (PVC), polystyrene, nylons, polyesters, acrylics, polyurethanes, and polycarbonates and also rubbers which have been around are natural rubber, styrene butadiene rubber, chloroprene, and butadiene rubber. Plastics are generally classified as *commodity*, *specialty* and *engineering* plastics.

$$\left[\begin{array}{c} CH_3 \\ | \\ CH-CH_2 \end{array}\right]_n$$

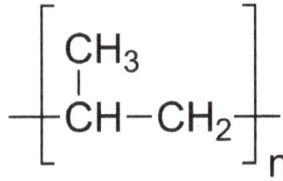

The repeating unit of the polymer polypropylene

Expanded polystyrene polymer packaging.

Polyvinyl chloride (PVC) is widely used, inexpensive, and annual production quantities are large. It lends itself to an vast array of applications, from artificial leather to electrical insulation and cabling, packaging, and containers. Its fabrication and processing are simple and well-established. The versatility of PVC is due to the wide range of plasticisers and other additives that it accepts. The term "additives" in polymer science refers to the chemicals and compounds added to the polymer base to modify its material properties.

Polycarbonate would be normally considered an engineering plastic (other examples include PEEK, ABS). Such plastics are valued for their superior strengths and other special material properties. They are usually not used for disposable applications, unlike commodity plastics.

Specialty plastics are materials with unique characteristics, such as ultra-high strength, electrical conductivity, electro-fluorescence, high thermal stability, etc.

The dividing lines between the various types of plastics is not based on material but rather on their properties and applications. For example, polyethylene (PE) is a cheap, low friction polymer commonly used to make disposable bags for shopping and trash, and is considered a commodity plastic, whereas medium-density polyethylene (MDPE) is used for underground gas and water pipes, and another variety called ultra-high-molecular-weight polyethylene (UHMWPE) is an engineering plastic which is used extensively as the glide rails for industrial equipment and the low-friction socket in implanted hip joints.

## Metal Alloys

The study of metal alloys is a significant part of materials science. Of all the metallic alloys in use today, the alloys of iron (steel, stainless steel, cast iron, tool steel, alloy steels) make up the largest proportion both by quantity and commercial value. Iron alloyed with various proportions of carbon gives low, mid and high carbon steels. An iron carbon alloy is only considered steel if the carbon level is between 0.01% and 2.00%. For the steels, the hardness and tensile strength of the steel is related to the amount of carbon present, with increasing carbon levels also leading to lower ductility and toughness. Heat treatment processes such as quenching and tempering can significantly change these properties however. Cast Iron is defined as an iron–carbon alloy with more than 2.00% but less than 6.67% carbon. Stainless steel is defined as a regular steel alloy with greater than 10% by weight alloying content of Chromium. Nickel and Molybdenum are typically also found in stainless steels.

Wire rope made from steel alloy.

Other significant metallic alloys are those of aluminium, titanium, copper and magnesium. Copper alloys have been known for a long time (since the Bronze Age), while the alloys of the other three metals have been relatively recently developed. Due to the chemical reactivity of these metals, the electrolytic extraction processes required were only developed relatively recently. The alloys of aluminium, titanium and magnesium are also known and valued for their high strength-to-weight ratios and, in the case of magnesium, their ability to provide electromagnetic shielding. These materials are ideal for situations where high strength-to-weight ratios are more important than bulk cost, such as in the aerospace industry and certain automotive engineering applications.

## Semiconductors

The study of semiconductors is a significant part of materials science. A semiconductor is a material that has a resistivity between a metal and insulator. It's electronic properties can be greatly altered through intentionally introducing impurities, or doping. From these semiconductor materials, things such as diodes, transistors, light-emitting diodes (LEDs), and analog and digital electric circuits can be built, making them materials of interest in industry. Semiconductor devices have replaced thermionic devices (vacuum tubes) in most applications. Semiconductor devices are manufactured both as single discrete devices and as integrated circuits (ICs), which consist of a number—from a few to millions—of devices manufactured and interconnected on a single semiconductor substrate.

Of all the semiconductors in use today, silicon makes up the largest portion both by quantity and commercial value. Monocrystalline silicon is used to produce wafers used in the semiconductor and electronics industry. Second to silicon, gallium arsenide (GaAs) is the second most popular semiconductor used. Due to its higher electron mobility and saturation velocity compared to silicon, its a material of choice for high speed electronics applications. These superior properties are compelling reasons to use GaAs circuitry in mobile phones, satellite communications, microwave point-to-point links and higher frequency radar systems. Other semiconductor materials include germanium, silicon carbide, and gallium nitride and have various applications.

## Relation to other Fields

Materials science evolved—starting from the 1960s—because it was recognized that to create, discover and design new materials, one had to approach it in a unified manner. Thus, materials science and engineering emerged at the intersection of various fields such as metallurgy, solid state physics, chemistry, chemical engineering, mechanical engineering and electrical engineering.

The field is inherently interdisciplinary, and the materials scientists/engineers must be aware and make use of the methods of the physicist, chemist and engineer. The field thus, maintains close relationships with these fields. Also, many physicists, chemists and engineers also find themselves working in materials science.

The overlap between physics and materials science has led to the offshoot field of *materials physics*, which is concerned with the physical properties of materials. The approach is generally more macroscopic and applied than in condensed matter physics.

The field of materials science and engineering is important both from a scientific perspective, as well as from an engineering one. When discovering new materials, one encounters new phenomena that may not have been observed before. Hence, there is lot of science to be discovered when working with materials. Materials science also provides test for theories in condensed matter physics.

Materials are of the utmost importance for engineers, as the usage of the appropriate materials is crucial when designing systems. As a result, materials science is an increasingly important part of an engineer's education.

# Emerging Technologies in Materials Science

| Emerging technology | Status | Potentially marginalized technologies | Potential applications | Related articles |
|---|---|---|---|---|
| Aerogel | Hypothetical, experiments, diffusion, early uses | Traditional insulation, glass | Improved insulation, insulative glass if it can be made clear, sleeves for oil pipelines, aerospace, high-heat & extreme cold applications | |
| Amorphous metal | Experiments | Kevlar | Armor | |
| Conductive polymers | Research, experiments, prototypes | Conductors | Lighter and cheaper wires, antistatic materials, organic solar cells | |
| Femtotechnology, pico-technology | Hypothetical | Present nuclear | New materials; nuclear weapons, power | |
| Fullerene | Experiments, diffusion | Synthetic diamond and carbon nanotubes (e.g., Buckypaper) | Programmable matter | |
| Graphene | Hypothetical, experiments, diffusion, early uses | Silicon-based integrated circuit | Components with higher strength to weight ratios, transistors that operate at higher frequency, lower cost of display screens in mobile devices, storing hydrogen for fuel cell powered cars, filtration systems, longer-lasting and faster-charging batteries, sensors to diagnose diseases | Potential applications of graphene |
| High-temperature superconductivity | Cryogenic receiver front-end (CRFE) RF and microwave filter systems for mobile phone base stations; prototypes in dry ice; Hypothetical and experiments for higher temperatures | Copper wire, semiconductor integral circuits | No loss conductors, frictionless bearings, magnetic levitation, lossless high-capacity accumulators, electric cars, heat-free integral circuits and processors | |

| Emerging technology | Status | Potentially marginalized technologies | Potential applications | Related articles |
|---|---|---|---|---|
| LiTraCon | Experiments, already used to make Europe Gate | Glass | Building skyscrapers, towers, and sculptures like Europe Gate | |
| Metamaterials | Hypothetical, experiments, diffusion | Classical optics | Microscopes, cameras, metamaterial cloaking, cloaking devices | |
| Metal foam | Research, commercialization | Hulls | Space colonies, floating cities | |
| Multi-function structures | Hypothetical, experiments, some prototypes, few commercial | Composite materials mostly | Wide range, e.g., self health monitoring, self healing material, morphing, ... | |
| Nanomaterials: carbon nanotubes | Hypothetical, experiments, diffusion, early uses | Structural steel and aluminium | Stronger, lighter materials, space elevator | Potential applications of carbon nanotubes, carbon fiber |
| Programmable matter | Hypothetical, experiments | Coatings, catalysts | Wide range, e.g., claytronics, synthetic biology | |
| Quantum dots | Research, experiments, prototypes | LCD, LED | Quantum dot laser, future use as programmable matter in display technologies (TV, projection), optical data communications (high-speed data transmission), medicine (laser scalpel) | |
| Silicene | Hypothetical, research | Field-effect transistors | | |
| Superalloy | Research, diffusion | Aluminum, titanium, composite materials | Aircraft jet engines | |
| Synthetic diamond | early uses (drill bits, jewelry) | Silicon transistors | Electronics | |

# Molecular Engineering

Molecular engineering is an emerging field of study concerned with the design and testing of molecular properties, behavior and interactions in order to assemble bet-ter materials, systems, and processes for specific functions. This approach, in which observable properties of a macroscopic system are influenced by direct alteration of a molecular structure, falls into the broader category of "bottom-up" design.

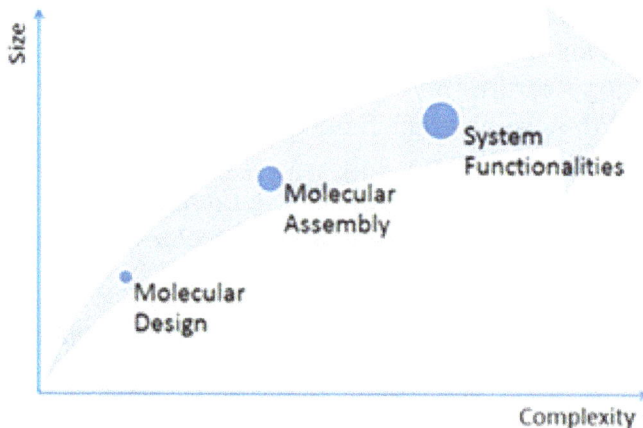

Molecular engineering deals with material development efforts in emerging technologies that require rigorous rational molecular design approaches towards systems of high complexity.

Molecular engineering is highly interdisciplinary by nature, encompassing aspects of chemical engineering, materials science, bioengineering, electrical engineering, phys-ics, mechanical engineering, and chemistry. There is also considerable overlap with nanotechnology, in that both are concerned with the behavior of materials on the scale of nanometers or smaller. Given the highly fundamental nature of molecular interac-tions, there are a plethora of potential application areas, limited perhaps only by one's imagination and the laws of physics. However, some of the early successes of molecular engineering have come in the fields of immunotherapy, synthetic biology, and printable electronics.

Molecular engineering is a dynamic and evolving field with complex target problems; breakthroughs require sophisticated and creative engineers who are conversant across disciplines. A rational engineering methodology that is based on molecular principles is in contrast to the widespread trial-and-error approaches common throughout engi-neering disciplines. Rather than relying on well-described but poorly-understood em-pirical correlations between the makeup of a system and its properties, a molecular design approach seeks to manipulate system properties directly using an understand-ing of their chemical and physical origins. This often gives rise to fundamentally new materials and systems, which are required to address outstanding needs in numerous fields, from energy to healthcare to electronics. Additionally, with the increased sophis-tication of technology, trial-and-error approaches are often costly and difficult, as it may be difficult to account for all relevant dependencies among variables in a complex system. Molecular engineering efforts may include computational tools, experimental methods, or a combination of both.

## History

Molecular engineering was first mentioned in the research literature in 1956 by Arthur R. von Hippel, who defined it as "... a new mode of thinking about engineering problems. Instead of taking prefabricated materials and trying to devise engineering applications consistent with their macroscopic properties, one builds materials from their atoms and molecules for the purpose at hand." This concept was echoed in Richard Feynman's seminal 1959 lecture *There's Plenty of Room at the Bottom*, which is widely regarded as giving birth to some of the fundamental ideas of the field of nanotechnology. In spite of the early introduction of these concepts, it was not until the mid-1980s with the publication of *Engines of Creation: The Coming Era of Nanotechnology* by Drexler that the modern concepts of nano and molecular-scale science began to grow in the public consciousness.

The discovery of electrically-conductive properties in polyacetylene by Alan J. Heeger in 1977 effectively opened the field of organic electronics, which has proved foundational for many molecular engineering efforts. Design and optimization of these materials has led to a number of innovations including organic light-emitting diodes and flexible solar cells.

## Applications

Molecular design has been an important element of many disciplines in academia, including bioengineering, chemical engineering, electrical engineering, materials science, mechanical engineering and chemistry. However, one of the on-going challenges is in bringing together the critical mass of manpower amongst disciplines to span the realm from design theory to materials production, and from device design to product development. Thus, while the concept of rational engineering of technology from the bottom-up is not new, it is still far from being widely translated into R&D efforts.

Molecular engineering is used in many industries. Some applications of technologies where molecular engineering plays a critical role:

### Consumer Products

- Antibiotic surfaces (eg. incorporation of silver nanoparticles or antibacterial peptides into coatings to prevent microbial infection)

- Cosmetics (eg. rheological modification with small molecules and surfactants in shampoo)

- Cleaning products (eg. nanosilver in laundry detergent)

- Consumer electronics (organic light-emitting diode displays (OLED))

- Electrochromic windows (eg. windows in Dreamliner 787)

- Zero emission vehicles (eg. advanced fuel cells/batteries)

- Self-cleaning surfaces (eg. super hydrophobic surface coatings)

## Energy Harvesting and Storage

- Flow batteries - Synthesizing molecules for high-energy density electrolytes and highly-selective membranes in grid-scale energy storage systems.

- Lithium-ion batteries - Creating new molecules for use as electrode binders, electrolytes, electrolyte additives, or even for energy storage directly in order to improve energy density (using materials such as graphene, silicon nanorods, and lithium metal), power density, cycle life, and safety.

- Solar cells - Developing new materials for more efficient and cost-effective solar cells including organic, quantum dot or perovskite-based photovoltaics.

- Photocatalytic water splitting - Enhancing the production of hydrogen fuel using solar energy and advanced catalytic materials such as semiconductor nanoparticles

## Environmental Engineering

- Water desalination (eg. new membranes for highly-efficient low-cost ion removal)

- Soil remediation (eg. catalytic nanoparticles that accelerate the degradation of long-lived soil contaminants such as chlorinated organic compounds)

- Carbon sequestration (eg. new materials for $CO_2$ adsorption)

## Immunotherapy

- Peptide-based vaccines (e.g. amphiphilic peptide macromolecular assemblies induce a robust immune response)

## Synthetic Biology

- CRISPR - Faster and more efficient gene editing technique

- Gene delivery/gene therapy - Designing molecules to deliver modified or new genes into cells of live organisms to cure genetic disorders

- Metabolic engineering - Modifying metabolism of organisms to optimize production of chemicals (eg. synthetic genomics)

- Protein engineering - Altering structure of existing proteins to enable specific new functions, or the creation of fully artificial proteins

## Techniques and Instruments used

Molecular engineers utilize sophisticated tools and instruments to make and analyze the interactions of molecules and the surfaces of materials at the molecular and nano-scale. The complexity of molecules being introduced at the surface is increasing, and the techniques used to analyze surface characteristics at the molecular level are ever-changing and improving. Meantime, advancements in high performance computing have greatly expanded the use of computer simulation in the study of molecular scale systems.

## Computational and Theoretical Approaches

- Computational chemistry
- High performance computing
- Molecular dynamics
- Molecular modeling
- Statistical mechanics
- Theoretical chemistry

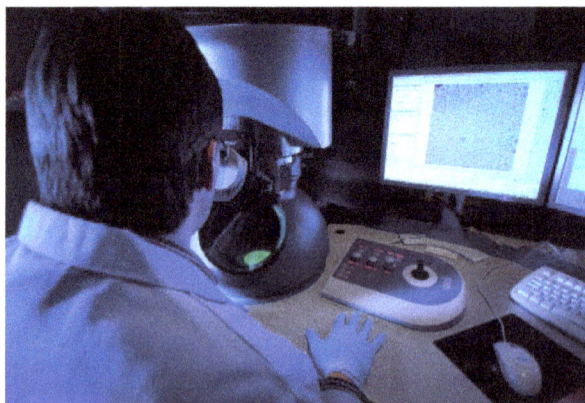

An EMSL scientist using the environmental transmission electron microscope at Pacific Northwest National Laboratory. The ETEM provides in situ capabilities that enable atomic-resolution imaging and spectroscopic studies of materials under dynamic operating conditions. In contrast to traditional operation of TEM under high vacuum, EMSL's ETEM uniquely allows imaging within high-temperature and gas environments.

## Microscopy

- Atomic Force Microscopy (AFM)
- Scanning Electron Microscopy (SEM)
- Transmission Electron Microscopy (TEM)

## Molecular Characterization

- Dynamic light scattering (DLS)
- Matrix-assisted laser desorption/ionization (MALDI) spectrocosopy
- Nuclear magnetic resonance (NMR) spectroscopy
- Size exclusion chromatography (SEC)

## Spectroscopy

- Ellipsometry
- 2D X-Ray Diffraction (XRD)
- Raman Spectroscopy/Microscopy

## Surface Science

- Glow Discharge Optical Emission Spectrometry
- Time of Flight-Secondary Ion Mass Spectrometry (ToF-SIMS)
- X-Ray Photoelectron Spectroscopy (XPS)

## Synthetic Methods

- DNA synthesis
- Nanoparticle synthesis
- Organic synthesis
- Peptide synthesis
- Polymer synthesis

## Other Tools

- Focused Ion Beam (FIB)
- Profilometer
- UV Photoelectron Spectroscopy (UPS)
- Vibrational Sum Frequency Generation

## Research / Education

At least three universities offer graduate degrees dedicated to molecular engineering:

the University of Chicago, the University of Washington, and Kyoto University. These programs are interdisciplinary institutes with faculty from several research areas.

The academic journal Molecular Systems Design & Engineering publishes research from a wide variety of subject areas that demonstrates "a molecular design or optimisation strategy targeting specific systems functionality and performance."

# Nanometrology

NIST Next-Generation Nanometrology research.

Nanometrology is a subfield of metrology, concerned with the science of measurement at the nanoscale level. Nanometrology has a crucial role in order to produce nanomaterials and devices with a high degree of accuracy and reliability in nanomanufacturing.

A challenge in this field is to develop or create new measurement techniques and standards to meet the needs of next-generation advanced manufacturing, which will rely on nanometer scale materials and technologies. The needs for measurement and characterization of new sample structures and characteristics far exceed the capabilities of current measurement science. Anticipated advances in emerging U.S. nanotechnology industries will require revolutionary metrology with higher resolution and accuracy than has previously been envisioned.

## Introduction

Control of the critical dimensions are the most important factors in nanotechnology. Nanometrology today, is to a large extent based on the development in semiconductor technology. Nanometrology is the science of measurement at the nanoscale level.

Nanometre or nm is equivalent to $10^{-9}$ m. In Nanotechnology accurate control of dimensions of objects is important. Typical dimensions of nanosystems vary from 10 nm to a few hundred nm and while fabricating such systems measurement up to 0.1 nm is required.

"Scanning Electron Microscope"

At nanoscale due to the small dimensions various new physical phenomena can be observed. For example, when the crystal size is smaller than the electron mean free path the conductivity of the crystal changes. Another example is the discretization of stresses in the system. It becomes important to measure the physical parameters so as to apply these phenomena into engineering of nanosystems and manufacturing them. The measurement of length or size, force, mass, electrical and other properties is included in Nanometrology. The problem is how to measure these with reliability and accuracy. The measurement techniques used for macro systems cannot be directly used for measurement of parameters in nanosystems. Various techniques based on physical phenomena have been developed which can be used for measure or determine the parameters for nanostructures and nanomaterials. Some of the popular ones are X-Ray Diffraction, Transmission Electron Microscopy, High Resolution Transmission Electron Microscopy, Atomic Force Microscopy, Scanning Electron Microscopy, Field Emission Scanning Electron Microscopy and Brunauer, Emmett, Teller method to determine specific surface.

Nanotechnology is an important field because of the large number of applications it has and it has become necessary to develop more precise techniques of measurement and globally accepted standards. Hence progress is required in the field of Nanometrology.

## Development Needs

Nanotechnology can be divided into two branches. The first being Molecular Nanotechnology which involves bottom up manufacturing and the second is Engineering Nanotechnology which involve the development and processing of materials and systems at nanoscale. The measurement and manufacturing tools and techniques required for the two branches are slightly different.

Furthermore, Nanometrology requirements are different for the industry and research institutions. Nanometrology of research has progressed faster than that for industry mainly because implementing nanometrology for industry is difficult. In research oriented nanometrology resolution is important whereas in industrial nanometrology accuracy is given precedence over resolution. Further due to economic reasons it is important to have low time costs in industrial nanometrology it is not important for research nanometrology. The various measurement techniques available today require a controlled environment like in vacuum, vibration and noise free environment. Also, in industrial nanometrology requires that the measurements be more quantitative with minimum number of parameters.

## Standards

### International Standards

Metrology standards are objects or ideas that are designated as being authoritative for some accepted reason. Whatever value they possess is useful for comparison to unknowns for the purpose of establishing or confirming an assigned value based on the standard. The execution of measurement comparisons for the purpose of establishing the relationship between a standard and some other measuring device is calibration. The ideal standard is independently reproducible without uncertainty. The world-wide market for products with nanotechnology applications is projected to be at least a couple of hundred billion dollars in the near future. Until recently, there almost no established internationally accepted standards for nanotechnology related field. The International Organisation for Standardization TC-229 Technical Committee on Nanotechnology recently published few standards for terminology, characterization of nanomaterials and nanoparticles using measurement tools like AFM, SEM, Interferometers, optoacoustic tools, gas adsorption methods etc. Certain standards for standardization of measurements for electrical properties have been published by the International Electrotechnical Commission. Some important standards which are yet to be established are standards for measuring thickness of thin films or layers, characterization of surface features, standards for force measurement at nanoscale, standards for characterization of critical dimensions of nanoparticles and nanostructures and also Standards for measurement for physical properties like conductivity, elasticity etc.

### National Standards

Because of the importance of nanotechnology in the future, countries around the world have programmes to establish national standards for nanometrology and nanotechnology. These programmes are run by the national standard agencies of the respective countries. In the United States, National Institute of Standards and Technology has been working on developing new techniques for measurement at nanoscale and has also established some national standards for nanotechnology. These standards are for nanoparticle characterization, Roughness Characterization, magnification standard, calibration standards etc.

## Calibration

It is difficult to provide samples using which precision instruments can be calibrated at nanoscale. Reference or Calibration standards are important for repeatability to be ensured. But there are no international standards for calibration and the calibration artefacts provided by the company along with their equipment is only good for calibrating that particular equipment. Hence it is difficult to select a universal calibration artefact using which we can achieve repeatability at nanoscale. At nanoscale while calibrating care needs to be taken for influence of external factors like vibration, noise, motions caused by thermal drift and creep and internal factors like the interaction between the artefact and the equipment which can cause significant deviations.

## Measurement Techniques

In the last 70 years various techniques for measuring at nanoscale have been developed. Most of them based on some physical phenomena observed on particle interactions or forces at nanoscale. Some of the most commonly used techniques are Atomic Force Microscopy, X-Ray Diffraction, Scanning Electron Microscopy, Transmission Electron Microscopy, High Resolution Transmission Electron Microscopy, and Field Emission Scanning Electron Microscopy.

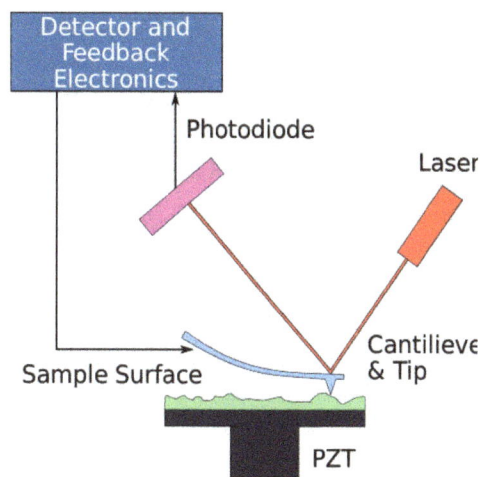

Block Diagram of *Atomic Force Microscope*.

Atomic Force Microscopy (AFM) is one of the most common measurement techniques. It can be used to measure Topology, grain size, frictional characteristics and different forces. It consists of a silicon cantilever with a sharp tip with a radius of curvature of a few nanometers. The tip is used as a probe on the specimen to be measured. The forces acting at the atomic level between the tip and the surface of the specimen cause the tip to deflect and this deflection is detected using a laser spot which is reflected to an array of photodiodes.

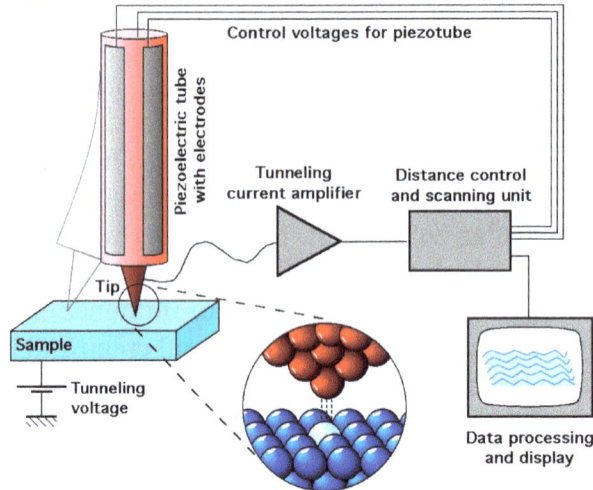

Diagram of *Scanning Tunneling Microscope.*

Scanning Tunneling Microscopy (STM) is another instrument commonly used. It is used to measure 3-D topology of the specimen. The STM is based on the concept of quantum tunneling. When a conducting tip is brought very near to the surface to be examined, a bias (voltage difference) applied between the two can allow electrons to tunnel through the vacuum between them. Measurements are made by monitoring the current as the tip's position scans across the surface, which can then be used to display an image.

Another commonly used instrument is the Scanning Electron Microscopy (SEM) which apart from measuring the shape and size of the particles and topography of the surface can be used to determine the composition of elements and compounds the sample is composed of. In SEM the specimen surface is scanned with a high energy electron beam. The electrons in the beam interact with atoms in the specimen and interactions are detected using detectors. The interactions produced are back scattering of electrons, transmission of electrons, secondary electrons etc. To remove high angle electrons magnetics lenses are used.

The instruments mentioned above produce realistic pictures of the surface are excellent measuring tools for research. Industrial applications of nanotechnology require the measurements to be produced need to be more quantitative. The requirement in industrial nanometrology is for higher accuracy than resolution as compared to research nanometrology.

# Nano Coordinate Measuring Machine

A Coordinate Measuring Machine (CMM) that works at the nanoscale would have a smaller frame than the CMM used for macroscale objects. This is so because it may provide the necessary stiffness and stability to achieve nanoscale uncertainties in x,y and z directions. The probes for such a machine need to be small to enable a 3-D measurement of nanometre features from the sides and from inside like nanoholes. Also for accuracy laser interferometers need to be used. NIST has developed a surface measuring instrument, called the Molecular Measuring Machine. This instrument is basically an STM. The x- and y-axes are read out by laser interferometers. The molecules on the surface area can be identified individually and at the same time the distance between any two molecules can be determined. For measuring with molecular resolution, the measuring times become very large for even a very small surface area. Ilmenau Machine is another nanomeasuring machine developed by researchers at the Ilmenau University of Technology.

Dimensional Metrology using *CMM*.

## Components of a Nano CMM

- Nanoprobes
- Control Hardware
- 3D-nanopositioning platform
- Instruments with high resolution and accuracy for linear measurement
- Instruments with high resolution and accuracy for angular measurement

## List of Some of the Measurement Techniques

1. Atomic Force Microscopy

2. X-ray absorption Spectroscopy

3. X- Ray Diffraction

4. Small Angle X-Ray Scattering

5. Scanning Tunneling Microscopy

6. Transmission Electron Microscopy

7. Capacitance Spectroscopy

8. Polarization Spectroscopy

9. Auger Electron Spectroscopy

10. Raman Spectroscopy

11. Small Angle Neutron Scattering

12. Scanning Electron Microscopy

13. Cyclic Voltammetry

14. Linear Sweep Voltammetry

15. Nuclear Magnetic Resonance

16. Mössbauer Spectroscopy

17. Fourier Transform Infrared Spectroscopy

18. Photoluminescence Spectroscopy

19. Electroluminescence Spectroscopy

20. Differential Scanning Calorimetry

21. Secondary Ion Mass Spectrometry

22. Cathodoluminescence Spectroscopy

23. Electron Energy Loss Spectroscopy

24. Energy Dispersive X-Ray Spectroscopy

25. Four point probe and I-V technique

26. X-Ray Photoelectron Spectroscopy

27. Scanning Near-field Optical Microscopy

28. Single-molecule Spectroscopy

29. Neutron Diffraction

30. Interference Microscopy

31. Laser Interferometry

## Traceability

In metrology at macro scale achieving traceability is quite easy and artefacts like scales, laser interferometers, step gauges, and straight edges are used. At nanoscale a crystalline highly oriented pyrolytic graphite (HOPG), mica or silicon surface is considered suitable used as calibration artefact for achieving traceability. But it is not always possible to ensure traceability. Like what is a straight edge at nanoscale and even if take the same standard as that for macroscale there is no way to calibrate it accurately at nanoscale. This so because the requisite internationally or nationally accepted reference standards are not always there. Also the measurement equipment required to ensure traceability has not been developed. The generally used for traceability are miniaturisation of traditional metrology standards hence there is a need for establishing nanoscale standards. Also there is a need to establish some kind of uncertainty estimation model. Traceability is one of the fundamental requirements for manufacturing and assembly of products when multiple producers are there.

## Tolerance

"Integrated Circuit" made using Monolithic Integration Technique.

Tolerance is the permissible limit or limits of variation in dimensions, properties, or conditions without significantly affecting functioning of equipment or a process. Tolerances are specified to allow reasonable leeway for imperfections and inherent variability without compromising performance. In nanotechnology the systems have dimensions in the range of nanometers. Defining tolerances at nanoscale with suitable calibration standards for traceability is difficult for different nanomanufacturing methods. There are various integration techniques developed in the semiconductor industry that are used in nanomanufacturing.

## Integration Techniques

- In Hetero integration direct fabrication of nanosystems from compound substrates is done. Geometric tolerances are required to achieve the functionality of the assembly.

- In Hybrid integration nanocomponents are placed or assembled on a substrate fabricating functioning nanosystems. In this technique, the most important control parameter is the positional accuracy of the components on the substrate.

- In Monolithic integration all the fabrication process steps are integrated on a single substrate and hence no mating of components or assembly is required. The advantage of this technique is that the geometric measurements are no longer of primary importance for achieving functionality of nanosystem or control of the fabrication process.

## Classification of Nanostructures

There are a variety of Nanostructures like nanocomposites, nanowires, nanopowders, nanotubes, fullerenes nanofibers, nanocages, nanocrystallites, nanoneedles, nanofoams, nanomeshes, nanoparticles, nanopillars, thin films, nanorods, nanofabrics, quantumdots etc. The most common way to classify nano structures is by their dimensions.

SEM of *Nanowire*.

## Dimensional Classification

| Dimensions | Criteria | Examples |
|---|---|---|
| Zero-dimensional (0-D) | The nanostructure has all dimensions in the nanometer range. | Nanoparticles, quantum dots, nanodots |
| One-dimensional (1-D) | One dimension of the nanostructure is outside the nanometer range. | Nanowires, nanorods, nanotubes |

| Two-dimensional (2-D) | Two dimensions of the nanostructure are outside the nanometer range. | Coatings, thin-film-multilayers |
| Three-dimensional (3-D) | Three dimensions of the nanostructure are outside the nanometer range. | Bulk |

## Classification of Grain Structure

Nanostructures can be classified on the basis of the grain structure and size there are made up of. This is applicable in the cas of 2-dimensional and 3-Dimensional Nanostructurs.

## Surface Area Measurement

For nanopowder to determine the specific surface area the B.E.T. method is commonly used. The drop of pressure of nitrogen in a closed container due to adsorption of the nitrogen molecules to the surface of the material inserted in the container is measured. Also, the shape of the nanopowder particles is assumed to be spherical.

$$D = 6/(\rho^*A)$$

Where "D" is the effective diameter, "$\rho$" is the density and "A" is the surface area fond from the B.E.T. method.

# Nanobiotechnology

Nanobiotechnology, bionanotechnology, and nanobiology are terms that refer to the intersection of nanotechnology and biology. Given that the subject is one that has only emerged very recently, bionanotechnology and nanobiotechnology serve as blanket terms for various related technologies.

This discipline helps to indicate the merger of biological research with various fields of nanotechnology. Concepts that are enhanced through nanobiology include: nanodevices (such as biological machines), nanoparticles, and nanoscale phenomena that occurs within the discipline of nanotechnology. This technical approach to biology allows scientists to imagine and create systems that can be used for biological research. Biologically inspired nanotechnology uses biological systems as the inspirations for technologies not yet created. However, as with nanotechnology and biotechnology, bionanotechnology does have many potential ethical issues associated with it.

The most important objectives that are frequently found in nanobiology involve applying nanotools to relevant medical/biological problems and refining these applications. Developing new tools, such as peptoid nanosheets, for medical and biological purposes is another primary objective in nanotechnology. New nanotools are often made by refining the applications of the nanotools that are already being used. The imaging of native biomolecules, biological membranes, and tissues is also a major topic for the nanobiology researchers. Other topics concerning nanobiology include the use of cantilever array sensors and the application of nanophotonics for manipulating molecular processes in living cells.

Recently, the use of microorganisms to synthesize functional nanoparticles has been of great interest. Microorganisms can change the oxidation state of metals. These microbial processes have opened up new opportunities for us to explore novel applications, for example, the biosynthesis of metal nanomaterials. In contrast to chemical and physical methods, microbial processes for synthesizing nanomaterials can be achieved in aqueous phase under gentle and environmentally benign conditions. This approach has become an attractive focus in current green bionanotechnology research towards sustainable development.

## Terminology

The terms are often used interchangeably. When a distinction is intended, though, it is based on whether the focus is on applying biological ideas or on studying biology with nanotechnology. Bionanotechnology generally refers to the study of how the goals of nanotechnology can be guided by studying how biological "machines" work and adapting these biological motifs into improving existing nanotechnologies or creating new ones. Nanobiotechnology, on the other hand, refers to the ways that nanotechnology is used to create devices to study biological systems.

In other words, nanobiotechnology is essentially miniaturized biotechnology, whereas bionanotechnology is a specific application of nanotechnology. For example, DNA nanotechnology or cellular engineering would be classified as bionanotechnology because they involve working with biomolecules on the nanoscale. Conversely, many new medical technologies involving nanoparticles as delivery systems or as sensors would be examples of nanobiotechnology since they involve using nanotechnology to advance the goals of biology.

The definitions enumerated above will be utilized whenever a distinction between nanobio and bionano is made in this article. However, given the overlapping usage of the terms in modern parlance, individual technologies may need to be evaluated to determine which term is more fitting. As such, they are best discussed in parallel.

## Concepts

Most of the scientific concepts in bionanotechnology are derived from other fields. Biochemical principles that are used to understand the material properties of biological systems are central in bionanotechnology because those same principles are to be used to create new technologies. Material properties and applications studied in bionanoscience include mechanical properties(e.g. deformation, adhesion, failure), electrical/electronic (e.g. electromechanical stimulation, capacitors, energy storage/batteries), optical (e.g. absorption, luminescence, photochemistry), thermal (e.g. thermomutability, thermal management), biological (e.g. how cells interact with nanomaterials, molecular flaws/defects, biosensing, biological mechanisms s.a. mechanosensing), nanoscience of disease (e.g. genetic disease, cancer, organ/tissue failure), as well as computing (e.g. DNA computing)and agriculture(target delivery of pesticides, hormones and fertilizers. The impact of bionanoscience, achieved through structural and mechanistic analyses of biological processes at nanoscale, is their translation into synthetic and technological applications through nanotechnology.

Nano-biotechnology takes most of its fundamentals from nanotechnology. Most of the devices designed for nano-biotechnological use are directly based on other existing nanotechnologies. Nano-biotechnology is often used to describe the overlapping multidisciplinary activities associated with biosensors, particularly where photonics, chemistry, biology, biophysics, nano-medicine, and engineering converge. Measurement in biology using wave guide techniques, such as dual polarization interferometry, are another example.

## Applications

Applications of bionanotechnology are extremely widespread. Insofar as the distinction holds, nanobiotechnology is much more commonplace in that it simply provides more tools for the study of biology. Bionanotechnology, on the other hand, promises to recreate biological mechanisms and pathways in a form that is useful in other ways.

## Nanomedicine

Nanomedicine is a field of medical science whose applications are increasing more and more thanks to nanorobots and biological machines, which constitute a very useful tool to develop this area of knowledge. In the past years, researchers have done many improvements in the different devices and systems required to develop nanorobots. This supposes a new way of treating and dealing with diseases such as cancer; thanks to nanorobots, side effects of chemotherapy have been controlled, reduced and even eliminated, so some years from now, cancer patients will be offered an alternative to treat this disease instead of chemotherapy, which causes secondary effects such as hair loss, fatigue or nausea killing not only cancerous cells but also the healthy ones. At a clinical level, cancer treatment with nanomedicine will consist on the supply of nanorobots to the patient through an injection that will seek for cancerous cells leaving untouched the healthy ones. Patients that will be treated through nanomedicine will not notice the presence of this nanomachines inside them; the only thing that is going to be noticeable is the progressive improvement of their health.

# Nanobiotechnology

Nanobiotechnology (sometimes referred to as nanobiology) is best described as help-ing modern medicine progress from treating symptoms to generating cures and regen-erating biological tissues. Three American patients have received whole cultured blad-ders with the help of doctors who use nanobiology techniques in their practice. Also, it has been demonstrated in animal studies that a uterus can be grown outside the body and then placed in the body in order to produce a baby. Stem cell treatments have been used to fix diseases that are found in the human heart and are in clinical trials in the United States. There is also funding for research into allowing people to have new limbs without having to resort to prosthesis. Artificial proteins might also become available to manufacture without the need for harsh chemicals and expensive machines. It has even been surmised that by the year 2055, computers may be made out of biochemicals and organic salts.

Another example of current nanobiotechnological research involves nanospheres coat-ed with fluorescent polymers. Researchers are seeking to design polymers whose flu-orescence is quenched when they encounter specific molecules. Different polymers would detect different metabolites. The polymer-coated spheres could become part of new biological assays, and the technology might someday lead to particles which could be introduced into the human body to track down metabolites associated with tumors and other health problems. Another example, from a different perspective, would be evaluation and therapy at the nanoscopic level, i.e. the treatment of Nanobacteria (25-200 nm sized) as is done by NanoBiotech Pharma.

While nanobiology is in its infancy, there are a lot of promising methods that will rely on nanobiology in the future. Biological systems are inherently nano in scale; nanoscience must merge with biology in order to deliver biomacromolecules and mo-lecular machines that are similar to nature. Controlling and mimicking the devices and processes that are constructed from molecules is a tremendous challenge to face the converging disciplines of nanotechnology. All living things, including humans, can be considered to be nanofoundries. Natural evolution has optimized the "natural" form of nanobiology over millions of years. In the 21st century, humans have devel-oped the technology to artificially tap into nanobiology. This process is best described as "organic merging with synthetic." Colonies of live neurons can live together on a biochip device; according to research from Dr. Gunther Gross at the University of North Texas. Self-assembling nanotubes have the ability to be used as a structural system. They would be composed together with rhodopsins; which would facilitate the optical computing process and help with the storage of biological materials. DNA (as the software for all living things) can be used as a structural proteomic system - a logical component for molecular computing. Ned Seeman - a researcher at New York University - along with other researchers are currently researching concepts that are similar to each other.

# Bionanotechnology

DNA nanotechnology is one important example of bionanotechnology. The utilization of the inherent properties of nucleic acids like DNA to create useful materials is a promising area of modern research. Another important area of research involves taking advantage of membrane properties to generate synthetic membranes. Proteins that self-assemble to generate functional materials could be used as a novel approach for the large-scale production of programmable nanomaterials. One example is the development of amyloids found in bacterial biofilms as engineered nanomaterials that can be programmed genetically to have different properties. Protein folding studies provide a third important avenue of research, but one that has been largely inhibited by our inability to predict protein folding with a sufficiently high degree of accuracy. Given the myriad uses that biological systems have for proteins, though, research into understanding protein folding is of high importance and could prove fruitful for bionanotechnology in the future.

Lipid nanotechnology is another major area of research in bionanotechnology, where physico-chemical properties of lipids such as their antifouling and self-assembly is exploited to build nanodevices with applications in medicine and engineering.

## Nanobiotechnology Applications in Agriculture

Meanwhile, nanotechnology application to biotechnology will also leave no field untouched by its groundbreaking scientific innovations for human wellness; the agricultural industry is no exception. Basically, nanomaterials are distinguished depending on the origin: natural, incidental and engineered nanoparticles. Among these, engineered nanoparticles have received wide attention in all fields of science, including medical, materials and agriculture technology with significant socio-economical growth. In the agriculture industry, engineered nanoparticles have been serving as nano carrier, containing herbicides, chemicals, or genes, which target particular plant parts to release their content. Previously nanocapsules containing herbicides have been reported to effectively penetrate through cuticles and tissues, allowing the slow and constant release of the active substances. Likewise, other literature describes that nano-encapsulated slow release of fertilizers has also become a trend to save fertilizer consumption and to minimize environmental pollution through precision farming. These are only a few examples from numerous research works which might open up exciting opportunities for nanobiotechnology application in agriculture. Also, application of this kind of engineered nanoparticles to plants should be considered the level of amicability before it is employed in agriculture practices. Based on a thorough literature survey, it was understood that there is only limited authentic information available to explain the biological consequence of engineered nanoparticles on treated plants. Certain reports underline the phytotoxicity of various origin of engineered nanoparticles to the plant caused by the subject of concentrations and sizes . At the same time, however, an equal number of studies were reported with a positive outcome of nanoparticles, which facilitate growth promoting nature to treat plant. In particular, compared to other nanoparticles, silver and gold nanoparticles based applications elicited beneficial results on various plant species with less and/or no toxicity. Silver nanoparticles (AgNPs) treated leaves of Asparagus showed the increased content of ascorbate and chlorophyll. Similarly, AgNPs-treated common bean and corn has increased shoot and root length, leaf surface area, chlorophyll, carbohydrate and protein contents reported earlier. The gold nanoparticle has been used to induce growth and seed yield in Brassica juncea.

## Tools

This field relies on a variety of research methods, including experimental tools (e.g. imaging, characterization via AFM/optical tweezers etc.), x-ray diffraction based tools, synthesis via self-assembly, characterization of self-assembly (using e.g. MP-SPR, DPI, recombinant DNA methods, etc.), theory (e.g. statistical mechanics, nanomechanics, etc.), as well as computational approaches (bottom-up multi-scale simulation, super-computing).

## References

- Ehud Gazit, Plenty of room for biology at the bottom: An introduction to bionanotechnology. Imperial College Press, 2007, ISBN 978-1-86094-677-6

- R. V. Lapshin (2016). "Drift-insensitive distributed calibration of probe microscope scanner in nanometer range: Virtual mode" (PDF). Applied Surface Science. Netherlands: Elsevier B. V. 378: 530–539. doi:10.1016/j.apsusc.2016.03.201. ISSN 0169-4332.

- Pelley, Janet. "Better Carbon Capture Through Chemistry | Chemical & Engineering News". cen.acs.org. Retrieved 2016-08-06.

- R. V. Lapshin (2015). "Drift-insensitive distributed calibration of probe microscope scanner in nanometer range: Approach description" (PDF). Applied Surface Science. Netherlands: Elsevier B. V. 359: 629–636. arXiv:1501.05545. Bibcode:2015ApSS..359..629L. doi:10.1016/j.apsusc.2015.10.108. ISSN 0169-4332.

- Alivisatos, Paul; Buchanan, Michelle (March 2010). Basic Research Needs for Carbon Capture: Beyond 2020 (Report). United States Department of Energy, Basic Energy Sciences Advisory Committee. Retrieved August 2015.

- Martin, Joseph D. (2015). "What's in a Name Change? Solid State Physics, Condensed Matter Physics, and Materials Science". Physics in Perspective. 17 (1): 3–32. Bibcode:2015PhP....17....3M. doi:10.1007/s00016-014-0151-7. Retrieved 20 April 2015.

- Tan, Shi; Ji, Ya J.; Zhang, Zhong R.; Yang, Yong (2014-07-21). "Recent Progress in Research on High-Voltage Electrolytes for Lithium-Ion Batteries". ChemPhysChem. 15 (10): 1956–1969. doi:10.1002/cphc.201402175. ISSN 1439-7641.

- Zhu, Ye; Li, Yan; Bettge, Martin; Abraham, Daniel P. (2012-01-01). "Positive Electrode Passivation by LiDFOB Electrolyte Additive in High-Capacity Lithium-Ion Cells". Journal of The Electrochemical Society. 159 (12): A2109–A2117. doi:10.1149/2.083212jes. ISSN 0013-4651.

- Black, Matthew; Trent, Amanda; Kostenko, Yulia; Lee, Joseph Saeyong; Olive, Colleen; Tirrell, Matthew (2012-07-24). "Self-Assembled Peptide Amphiphile Micelles Containing a Cytotoxic

T-Cell Epitope Promote a Protective Immune Response In Vivo". Advanced Materials. 24 (28): 3845–3849. doi:10.1002/adma.201200209. ISSN 1521-4095.

- "Pentagon Developing Shape-Shifting 'Transformers' for Battlefield". Fox News. 10 June 2009. Retrieved 26 April 2011.

- "IBM Research | IBM Research | Silicon Integrated Nanophotonics". Domino.research.ibm.com. 2010-03-04. Retrieved 2010-03-15.

# 7

# Evolution of Nanoengineering

Nanoengineering has evolved over a period of years. It traces the growth of the ideas and concepts falling under the broad category of nanoengineering. The content within the chapter helps the reader in developing an in-depth understanding of the history of nanoengineering and nanotechnology.

The history of nanotechnology traces the development of the concepts and experimental work falling under the broad category of nanotechnology. Although nanotechnology is a relatively recent development in scientific research, the development of its central concepts happened over a longer period of time. The emergence of nanotechnology in the 1980s was caused by the convergence of experimental advances such as the invention of the scanning tunneling microscope in 1981 and the discovery of fullerenes in 1985, with the elucidation and popularization of a conceptual framework for the goals of nanotechnology beginning with the 1986 publication of the book *Engines of Creation*. The field was subject to growing public awareness and controversy in the early 2000s, with prominent debates about both its potential implications as well as the feasibility of the applications envisioned by advocates of molecular nanotechnology, and with governments moving to promote and fund research into nanotechnology. The early 2000s also saw the beginnings of commercial applications of nanotechnology, although these were limited to bulk applications of nanomaterials rather than the transformative applications envisioned by the field.

## Conceptual Origins

## Richard Feynman

Richard Feynman gave a 1959 talk which many years later inspired the conceptual foundations of nanotechnology.

The American physicist Richard Feynman lectured, "There's Plenty of Room at the Bottom," at an American Physical Society meeting at Caltech on December 29, 1959, which is often held to have provided inspiration for the field of nanotechnology. Feynman had described a process by which the ability to manipulate individual atoms and molecules might be developed, using one set of precise tools to build and operate another proportionally smaller set, so on down to the needed scale. In the course of this, he noted, scaling issues would arise from the changing magnitude of various physical phenomena: gravity would become less important, surface tension and Van der Waals attraction would become more important.

After Feynman's death, scholars studying the historical development of nanotechnology have concluded that his actual role in catalyzing nanotechnology research was limited, based on recollections from many of the people active in the nascent field in the 1980s and 1990s. Chris Toumey, a cultural anthropologist at the University of South Carolina, found that the published versions of Feynman's talk had a negligible influence in the twenty years after it was first published, as measured by citations in the scientific literature, and not much more influence in the decade after the Scanning Tunneling Microscope was invented in 1981. Subsequently, interest in "Plenty of Room" in the scientific literature greatly increased in the early 1990s. This is probably because the term "nanotechnology" gained serious attention just before that time, following its use by K. Eric Drexler in his 1986 book, *Engines of Creation: The Coming Era of Nanotechnology*, which took the Feynman concept of a billion tiny factories and added the idea that they could make more copies of themselves via computer control instead of control by a human operator; and in a cover article headlined "Nanotechnology", published later that year in a mass-circulation science-oriented magazine, *OMNI*. Toumey's analysis also includes comments from distinguished scientists in nanotechnology who say that "Plenty of Room" did not influence their early work, and in fact most of them had not read it until a later date.

These and other developments hint that the retroactive rediscovery of Feynman's "Plenty of Room" gave nanotechnology a packaged history that provided an early date of December 1959, plus a connection to the charisma and genius of Richard Feynman. Feynman's stature as a Nobel laureate and as an iconic figure in 20th century science surely helped advocates of nanotechnology and provided a valuable intellectual link to the past.

## Norio Taniguchi

The Japanese scientist called Norio Taniguchi of the Tokyo University of Science was the first to use the term "nano-technology" in a 1974 conference, to describe semiconductor processes such as thin film deposition and ion beam milling exhibiting characteristic control on the order of a nanometer. His definition was, "'Nano-technology' mainly consists of the processing of, separation, consolidation, and deformation of materials by one atom or one molecule." However, the term was not used again until 1981 when Eric Drexler, who was unaware of Taniguchi's prior use of the term, published his first paper on nanotechnology in 1981.

# K. Eric Drexler

K. Eric Drexler developed and popularized the concept of nanotechnology and founded the field of molecular nanotechnology.

In the 1980s the idea of nanotechnology as a deterministic, rather than stochastic, handling of individual atoms and molecules was conceptually explored in depth by K. Eric Drexler, who promoted the technological significance of nano-scale phenomena and devices through speeches and two influential books.

In 1980, Drexler encountered Feynman's provocative 1959 talk "There's Plenty of Room at the Bottom" while preparing his initial scientific paper on the subject, "Molecular Engineering: An approach to the development of general capabilities for molecular manipulation," published in the *Proceedings of the National Academy of Sciences* in 1981. The term "nanotechnology" (which paralleled Taniguchi's "nano-technology") was independently applied by Drexler in his 1986 book *Engines of Creation: The Coming Era of Nanotechnology*, which proposed the idea of a nanoscale "assembler" which would be able to build a copy of itself and of other items of arbitrary complexity. He also first published the term "grey goo" to describe what might happen if a hypothetical self-replicating machine, capable of independent operation, were constructed and released. Drexler's vision of nanotechnology is often called "Molecular Nanotechnology" (MNT) or "molecular manufacturing."

His 1991 Ph.D. work at the MIT Media Lab was the first doctoral degree on the topic of molecular nanotechnology and (after some editing) his thesis, "Molecular Machinery and Manufacturing with Applications to Computation," was published as *Nanosystems: Molecular Machinery, Manufacturing, and Computation,* which received the Association of American Publishers award for Best Computer Science Book of 1992. Drexler founded the Foresight Institute in 1986 with the mission of "Preparing for nanotechnology." Drexler is no longer a member of the Foresight Institute.

## Experimental Advances

Nanotechnology and nanoscience got a boost in the early 1980s with two major developments: the birth of cluster science and the invention of the scanning tunneling microscope (STM). These developments led to the discovery of fullerenes in 1985 and the structural assignment of carbon nanotubes a few years later

## Invention of Scanning Probe Microscopy

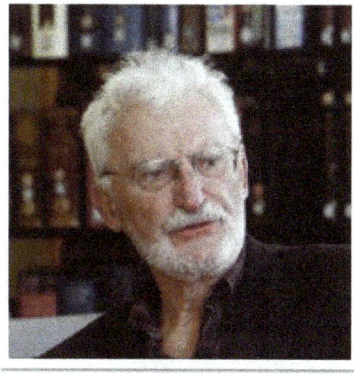

Gerd Binnig (left) and Heinrich Rohrer (right) won the 1986 Nobel Prize in Physics
for their 1981 invention of the scanning tunneling microscope.

The scanning tunneling microscope, an instrument for imaging surfaces at the atomic level, was developed in 1981 by Gerd Binnig and Heinrich Rohrer at IBM Zurich Research Laboratory, for which they were awarded the Nobel Prize in Physics in 1986. Binnig, Calvin Quate and Christoph Gerber invented the first atomic force microscope in 1986. The first commercially available atomic force microscope was introduced in 1989.

IBM researcher Don Eigler was the first to manipulate atoms using a scanning tunneling microscope in 1989. He used 35 Xenon atoms to spell out the IBM logo. He shared the 2010 Kavli Prize in Nanoscience for this work.

## Advances in Interface and Colloid Science

Interface and colloid science had existed for nearly a century before they became associated with nanotechnology. The first observations and size measurements of nanoparticles had been made during the first decade of the 20th century by Richard Adolf Zsigmondy, winner of the 1925 Nobel Prize in Chemistry, who made a detailed study of gold sols and other nanomaterials with sizes down to 10 nm using an ultramicroscope which was capable of visualizing particles much smaller than the light wavelength. Zsigmondy was also the first to use the term "nanometer" explicitly for characterizing particle size. In the 1920s, Irving Langmuir, winner of the 1932 Nobel Prize in Chemistry, and Katharine B. Blodgett introduced the concept of a monolayer, a layer of material one molecule thick. In the early 1950s, Derjaguin and Abrikosova conducted the first measurement of surface forces.

In 1974 the process of atomic layer deposition for depositing uniform thin films one atomic layer at a time was developed and patented by Tuomo Suntola and co-workers in Finland.

In another development, the synthesis and properties of semiconductor nanocrystals were studied. This led to a fast increasing number of semiconductor nanoparticles of quantum dots.

## Discovery of Fullerenes

Harry Kroto (left) won the 1996 Nobel Prize in Chemistry along with Richard Smalley (pictured below) and Robert Curl for their 1985 discovery of buckminsterfullerene, while Sumio Iijima (right) won the inaugural 2008 Kavli Prize in Nanoscience for his 1991 discovery of carbon nanotubes.

Fullerenes were discovered in 1985 by Harry Kroto, Richard Smalley, and Robert Curl, who together won the 1996 Nobel Prize in Chemistry. Smalley's research in physical chemistry investigated formation of inorganic and semiconductor clusters using pulsed molecular beams and time of flight mass spectrometry. As a consequence of this expertise, Curl introduced him to Kroto in order to investigate a question about the constituents of astronomical dust. These are carbon rich grains expelled by old stars such as R Corona Borealis. The result of this collaboration was the discovery of $C_{60}$ and the fullerenes as the third allotropic form of carbon. Subsequent discoveries included the endohedral fullerenes, and the larger family of fullerenes the following year.

The discovery of carbon nanotubes is largely attributed to Sumio Iijima of NEC in 1991, although carbon nanotubes have been produced and observed under a variety of conditions prior to 1991. Iijima's discovery of multi-walled carbon nanotubes in the insoluble material of arc-burned graphite rods in 1991 and Mintmire, Dunlap, and White's independent prediction that if single-walled carbon nanotubes could be made, then they would exhibit remarkable conducting properties  helped create the initial buzz that is now associated with carbon nanotubes. Nanotube research accelerated greatly following the independent discoveries by Bethune at IBM and Iijima at NEC of *single-walled* carbon nanotubes and methods to specifically produce them by adding transition-metal catalysts to the carbon in an arc discharge.

In the early 1990s Huffman and Kraetschmer, of the University of Arizona, discovered how to synthesize and purify large quantities of fullerenes. This opened the door to their characterization and functionalization by hundreds of investigators in government and industrial laboratories. Shortly after, rubidium doped $C_{60}$ was found to be a mid temperature (Tc = 32 K) superconductor. At a meeting of the Materials Research Society in 1992, Dr. T. Ebbesen (NEC) described to a spellbound audience his discovery and characterization of carbon nanotubes. This event sent those in attendance and others downwind of his presentation into their laboratories to reproduce and push those discoveries forward. Using the same or similar tools as those used by Huffman and Kratschmer, hundreds of researchers further developed the field of nanotube-based nanotechnology.

## Government Support

## National Nanotechnology Initiative

Mihail Roco of the National Science Foundation formally proposed the National Nanotechnology Initiative to the White House, and was a key architect in its initial development.

The National Nanotechnology Initiative is a United States federal nanotechnology research and development program. "The NNI serves as the central point of communication, cooperation, and collaboration for all Federal agencies engaged in nanotechnology research, bringing together the expertise needed to advance this broad and complex field." Its goals are to advance a world-class nanotechnology research and development (R&D) program, foster the transfer of new technologies into products for commercial and public benefit, develop and sustain educational resources, a skilled workforce, and the supporting infrastructure and tools to advance nanotechnology, and support responsible development of nanotechnology. The initiative was spearheaded by Mihail Roco, who formally proposed the National Nanotechnology Initiative to the Office of Science and Technology Policy during the Clinton administration in 1999, and was a key architect in its development. He is currently the Senior Advisor for Nanotechnology at the National Science Foundation, as well as the founding chair of the National Science and Technology Council subcommittee on Nanoscale Science, Engineering and Technology.

President Bill Clinton advocated nanotechnology development. In a 21 January 2000 speech at the California Institute of Technology, Clinton said, "Some of our research goals may take twenty or more years to achieve, but that is precisely why there is an important role for the federal government." Feynman's stature and concept of atomically precise fabrication played a role in securing funding for nanotechnology research, as mentioned in President Clinton's speech:

My budget supports a major new National Nanotechnology Initiative, worth $500 million. Caltech is no stranger to the idea of nanotechnology the ability to manipulate matter at the atomic and molecular level. Over 40 years ago, Caltech's own Richard Feynman asked, "What would happen if we could arrange the atoms one by one the way we want them?"

President George W. Bush further increased funding for nanotechnology. On December 3, 2003 Bush signed into law the 21st Century Nanotechnology Research and Development Act, which authorizes expenditures for five of the participating agencies totaling US$3.63 billion over four years. The NNI budget supplement for Fiscal Year 2009 provides $1.5 billion to the NNI, reflecting steady growth in the nanotechnology investment.

## Growing public Awareness and Controversy

### "Why the Future doesn't Need us"

"Why the future doesn't need us" is an article written by Bill Joy, then Chief Scientist at Sun Microsystems, in the April 2000 issue of *Wired* magazine. In the article, he argues that "Our most powerful 21st-century technologies — robotics, genetic engineering, and nanotech — are threatening to make humans an endangered species." Joy argues that developing technologies provide a much greater danger to humanity than any technology before it has ever presented. In particular, he focuses on genetics, nanotechnology and robotics. He argues that 20th-century technologies of destruction, such as the nuclear bomb, were limited to large governments, due to the complexity and cost of such devices, as well as the difficulty in acquiring the required materials. He also voices concern about increasing computer power. His worry is that computers will eventually become more intelligent than we are, leading to such dystopian scenarios as robot rebellion. He notably quotes the Unabomber on this topic. After the publication of the article, Bill Joy suggested assessing technologies to gauge their implicit dangers, as well as having scientists refuse to work on technologies that have the potential to cause harm.

In the AAAS Science and Technology Policy Yearbook 2001 article titled *A Response to Bill Joy and the Doom-and-Gloom Technofuturists*, Bill Joy was criticized for having technological tunnel vision on his prediction, by failing to consider social factors. In Ray Kurzweil's *The Singularity Is Near*, he questioned the regulation of potentially dangerous technology, asking "Should we tell the millions of people afflicted with cancer and other devastating conditions that we are canceling the development of all bioengineered treatments because there is a risk that these same technologies may someday be used for malevolent purposes?".

## Prey

*Prey* is a 2002 novel by Michael Crichton which features an artificial swarm of nanorobots which develop intelligence and threaten their human inventors. The novel generated concern within the nanotechnology community that the novel could negatively affect public perception of nanotechnology by creating fear of a similar scenario in real life.

## Drexler–smalley Debate

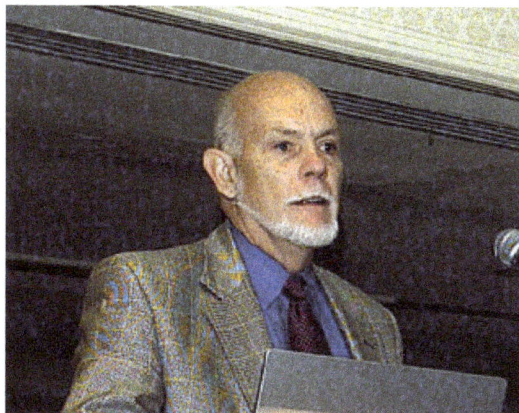

Richard Smalley, a co-discoverer of the fullerenes, was involved in a public debate with Eric Drexler about the feasibility of molecular assemblers.

Richard Smalley, best known for co-discovering the soccer ball-shaped "buckyball" molecule and a leading advocate of nanotechnology and its many applications, was an outspoken critic of the idea of molecular assemblers, as advocated by Eric Drexler. In 2001 he introduced scientific objections to them attacking the notion of universal assemblers in a 2001 *Scientific American* article, leading to a rebuttal later that year from Drexler and colleagues, and eventually to an exchange of open letters in 2003.

Smalley criticized Drexler's work on nanotechnology as naive, arguing that chemistry is extremely complicated, reactions are hard to control, and that a universal assembler is science fiction. Smalley believed that such assemblers were not physically possible and introduced scientific objections to them. His two principal technical objections, which he had termed the "fat fingers problem" and the "sticky fingers problem", argued against the feasibility of molecular assemblers being able to precisely select and place individual atoms. He also believed that Drexler's speculations about apocalyptic dangers of molecular assemblers threaten the public support for development of nanotechnology.

Smalley first argued that "fat fingers" made MNT impossible. He later argued that nanomachines would have to resemble chemical enzymes more than Drexler's assemblers and could only work in water. He believed these would exclude the possibility of "molecular assemblers" that worked by precision picking and placing of individual atoms.

Also, Smalley argued that nearly all of modern chemistry involves reactions that take place in a solvent (usually water), because the small molecules of a solvent contribute many things, such as lowering binding energies for transition states. Since nearly all known chemistry requires a solvent, Smalley felt that Drexler's proposal to use a high vacuum environment was not feasible.

Smalley also believed that Drexler's speculations about apocalyptic dangers of self-replicating machines that have been equated with "molecular assemblers" would threaten the public support for development of nanotechnology. To address the debate between Drexler and Smalley regarding molecular assemblers *Chemical & Engineering News* published a point-counterpoint consisting of an exchange of letters that addressed the issues.

Drexler and coworkers responded to these two issues in a 2001 publication. Drexler and colleagues noted that Drexler never proposed universal assemblers able to make absolutely anything, but instead proposed more limited assemblers able to make a very wide variety of things. They challenged the relevance of Smalley's arguments to the more specific proposals advanced in *Nanosystems*. Drexler maintained that both were straw man arguments, and in the case of enzymes, Prof. Klibanov wrote in 1994, "...using an enzyme in organic solvents eliminates several obstacles..." Drexler also addresses this in Nanosystems by showing mathematically that well designed catalysts can provide the effects of a solvent and can fundamentally be made even more efficient than a solvent/enzyme reaction could ever be. Drexler had difficulty in getting Smalley to respond, but in December 2003, *Chemical & Engineering News* carried a 4 part debate.

Ray Kurzweil spends four pages in his book 'The Singularity Is Near' to showing that Richard Smalley's arguments are not valid, and disputing them point by point. Kurzweil ends by stating that Drexler's visions are very practicable and even happening already.

## Royal Society Report on the Implications of Nanotechnology

The Royal Society and Royal Academy of Engineering's 2004 report on the implications of nanoscience and nanotechnologies was inspired by Prince Charles' concerns about nanotechnology, including molecular manufacturing. However, the report spent almost no time on molecular manufacturing. In fact, the word "Drexler" appears only once in the body of the report (in passing), and "molecular manufacturing" or "molecular nanotechnology" not at all. The report covers various risks of nanoscale technologies, such as nanoparticle toxicology. It also provides a useful overview of several nanoscale fields. The report contains an annex (appendix) on grey goo, which cites a weaker variation of Richard Smalley's contested argument against molecular manufacturing. It concludes that there is no evidence that autonomous, self replicating nanomachines will be developed in the foreseeable future, and suggests that regulators should be more concerned with issues of nanoparticle toxicology.

## Initial Commercial Applications

The early 2000s saw the beginnings of the use of nanotechnology in commercial products, although most applications are limited to the bulk use of passive nanomaterials. Examples include titanium dioxide and zinc oxide nanoparticles in sunscreen, cosmetics and some food products; silver nanoparticles in food packaging, clothing, disinfectants and household appliances such as Silver Nano; carbon nanotubes for stain-resistant textiles; and cerium oxide as a fuel catalyst. As of March 10, 2011, the Project on Emerging Nanotechnologies estimated that over 1300 manufacturer-identified nanotech products are publicly available, with new ones hitting the market at a pace of 3–4 per week.

The National Science Foundation funded researcher David Berube to study the field of nanotechnology. His findings are published in the monograph Nano-Hype: The Truth Behind the Nanotechnology Buzz. This study concludes that much of what is sold as "nanotechnology" is in fact a recasting of straightforward materials science, which is leading to a "nanotech industry built solely on selling nanotubes, nanowires, and the like" which will "end up with a few suppliers selling low margin products in huge volumes." Further applications which require actual manipulation or arrangement of nanoscale components await further research. Though technologies branded with the term 'nano' are sometimes little related to and fall far short of the most ambitious and transformative technological goals of the sort in molecular manufacturing proposals, the term still connotes such ideas. According to Berube, there may be a danger that a "nano bubble" will form, or is forming already, from the use of the term by scientists and entrepreneurs to garner funding, regardless of interest in the transformative possibilities of more ambitious and far-sighted work.

## References

- Gribbin, John; Gribbin, Mary (1997). Richard Feynman: A Life in Science. Dutton. p. 170. ISBN 0-452-27631-4.

- Bassett, Deborah R. (2010). "Taniguchi, Norio". In Guston, David H. Encyclopedia of nanoscience and society. London: SAGE. p. 747. ISBN 9781452266176. Retrieved 3 August 2014.

- Drexler, K. Eric (1992). Nanosystems: Molecular Machinery, Manufacturing, and Computation. Wiley. ISBN 0-471-57518-6. Retrieved 14 May 2011.

- Milburn, Colin (2008). Nanovision: Engineering the Future. Duke University Press. ISBN 0-8223-4265-0.

- Shankland, Stephen (28 September 2009). "IBM's 35 atoms and the rise of nanotech". CNET. Retrieved 12 May 2011.

- "President Clinton's Address to Caltech on Science and Technology". California Institute of Technology. Retrieved 13 May 2011.

- Jones, Richard M. (21 January 2000). "President Requests Significant Increase in FY 2001 Research Budget". FYI: The AIP Bulletin of Science Policy News. American Institute of Physics. Retrieved 13 May 2011.

- "Brown, John Seely and Duguid, Paul" (13 April 2000). "A Response to Bill Joy and the Doom-and-Gloom Technofuturists" (PDF). Retrieved 12 May 2011.

- "Nanoscience and nanotechnologies: opportunities and uncertainties". Royal Society and Royal Academy of Engineering. July 2004. Retrieved 13 May 2011.

- "Nanotechnology Information Center: Properties, Applications, Research, and Safety Guidelines". American Elements. Retrieved 13 May 2011.

- "Analysis: This is the first publicly available on-line inventory of nanotechnology-based consumer products". The Project on Emerging Nanotechnologies. 2008. Retrieved 13 May 2011.

- Phoenix, Chris (December 2003). "Of Chemistry, Nanobots, and Policy". Center for Responsible Nanotechnology. Retrieved 12 May 2011.

- Drexler, K. Eric; Forrest, David; Freitas, Robert A.; Hall, J. Storrs; Jacobstein, Neil; McKendree, Tom; Merkle, Ralph; Peterson, Christine (2001). "Debate About Assemblers — Smalley Rebuttal". Institute for Molecular Manufacturing. Retrieved 9 May 2010.

- "Nanotechnology: Drexler and Smalley make the case for and against 'molecular assemblers'". Chemical & Engineering News. American Chemical Society. 81 (48): 37–42. 1 December 2003. doi:10.1021/cen-v081n036.p037. Retrieved 9 May 2010.

# Permissions

# Index

www.ingramcontent.com/pod-product-compliance
Lightning Source LLC
Chambersburg PA
CBHW070150240326

41458CB00126B/2746